UNVEILING LARGE-SCALE STRUCTURES BEHIND THE MILKY WAY

A SERIES OF BOOKS ON RECENT DEVELOPMENTS IN ASTRONOMY AND ASTROPHYSICS

© Copyright 1994 Astronomical Society of the Pacific
390 Ashton Avenue, San Francisco, California 94112

Printed by BookCrafters, Inc.

First published 1994

Library of Congress Catalog Card Number: 94-78716
ISBN 0-937707-86-4

D. Harold McNamara, Managing Editor of Conference Series
408 ESC Brigham Young University
Provo, UT 84602
801-378-2298

A SERIES OF BOOKS ON RECENT DEVELOPMENTS IN ASTRONOMY AND ASTROPHYSICS

Vol. 19-Radio Interferometry: Theory, Techniques, and Applications, IAU Colloquium 131
ed. T. J. Cornwell and R. A. Perley   ISBN 0-937707-38-4

Vol. 20-Frontiers of Stellar Evolution, celebrating the 50th Anniversary of McDonald Observatory
ed. D. L. Lambert   ISBN 0-937707-39-2

Vol. 21-The Space Distribution of Quasars
ed. D. Crampton   ISBN 0-937707-40-6

Vol. 22-Nonisotropic and Variable Outflows from Stars
ed. L. Drissen, C. Leitherer, and A. Nota   ISBN 0-937707-41-4

Vol. 23-Astronomical CCD Observing and Reduction Techniques
ed. S. B. Howell   ISBN 0-937707-42-4

Vol. 24-Cosmology and Large-Scale Structure in the Universe
ed. R. R. de Carvalho   ISBN 0-937707-43-0

Vol. 25-Astronomical Data Analysis Software and Systems I
ed. D. M. Worrall, C. Biemesderfer, and J. Barnes   ISBN 0-937707-44-9

Vol. 26-Cool Stars, Stellar Systems, and the Sun, Seventh Cambridge Workshop
ed. M. S. Giampapa and J. A. Bookbinder   ISBN 0-937707-45-7

Vol. 27-The Solar Cycle
ed. K. L. Harvey   ISBN 0-937707-46-5

Vol. 28-Automated Telescopes for Photometry and Imaging
ed. S. J. Adelman, R. J. Dukes, Jr., and C. J. Adelman ISBN 0-937707-47-3

Vol. 29-Workshop on Cataclysmic Variable Stars
ed. N. Vogt   ISBN 0-937707-48-1

Vol. 30-Variable Stars and Galaxies, in honor of M. S. Feast on his retirement
ed. B. Warner   ISBN 0-937707-49-X

Vol. 31-Relationships Between Active Galactic Nuclei and Starburst Galaxies
ed. A. V. Filippenko   ISBN 0-937707-50-3

Vol. 32-Complementary Approaches to Double and Multiple Star Research, IAU Colloquium 135
ed. H. A. McAlister and W. I. Hartkopf   ISBN 0-937707-51-1

Vol. 33-Research Amateur Astronomy
ed. S. J. Edberg   ISBN 0-937707-52-X

Vol. 34-Robotic Telescopes in the 1990s
ed. A. V. Filippenko   ISBN 0-937707-53-8

Vol. 35-Massive Stars: Their Lives in the Interstellar Medium
ed. J. P. Cassinelli and E. B. Churchwell   ISBN 0-937707-54-6

Vol. 36-Planets and Pulsars
ed. J. A. Phillips, S. E. Thorsett, and S. R. Kulkarni ISBN 0-937707-55-4

Vol. 37-Fiber Optics in Astronomy II
ed. P. M. Gray   ISBN 0-937707-56-2

Vol. 38-New Frontiers in Binary Star Research
ed. K. C. Leung and I. S. Nha   ISBN 0-937707-57-0

Vol. 39-The Minnesota Lectures on the Structure and Dynamics of the Milky Way
ed. Roberta M. Humphreys   ISBN 0-937707-58-9

Vol. 40-Inside the Stars, IAU Colloquium 137
ed. Werner W. Weiss and Annie Baglin   ISBN 0-937707-59-7

Vol. 41-Astronomical Infrared Spectroscopy: Future Observational Directions
ed. Sun Kwok   ISBN 0-937707-60-0

Vol. 42-GONG 1992: Seismic Investigation of the Sun and Stars
ed. Timothy M. Brown   ISBN 0-937707-61-9

Vol. 43-Sky Surveys: Protostars to Protogalaxies
ed. B. T. Soifer   ISBN 0-937707-62-7

Vol. 44-Peculiar Versus Normal Phenomena in A-Type and Related Stars
ed. M. M. Dworetsky, F. Castelli, and R. Faraggiana   ISBN 0-937707-63-5

Vol. 45-Luminous High-Latitude Stars
ed. D. D. Sasselov   ISBN 0-937707-64-3

Vol. 46-The Magnetic and Velocity Fields of Solar Active Regions, IAU Colloquium 141
ed. H. Zirin, G. Ai, and H. Wang   ISBN 0-937707-65-1

Vol. 47-Third Decinnial US-USSR Conference on SETI
ed. G. Seth Shostak   ISBN 0-937707-66-X

Vol. 48-The Globular Cluster-Galaxy Connection
ed. Graeme H. Smith and Jean P. Brodie   ISBN 0-937707-67-8

Vol. 49-Galaxy Evolution: The Milky Way Perspective
ed. Steven R. Majewski   ISBN 0-937707-68-6

Vol. 50-Structure and Dynamics of Globular Clusters
ed. S. G. Djorgovski and G. Meylan   ISBN 0-937707-69-4

Vol. 51-Observational Cosmology
ed. G. Chincarini, A. Iovino, T. Maccacaro, and D. Maccagni   ISBN 0-937707-70-8

Vol. 52-Astronomical Data Analysis Software and Systems II
ed. R. J. Hanisch, J. V. Brissenden, and Jeannette Barnes   ISBN 0-937707-71-6

Vol. 53-Blue Stragglers
ed. Rex A. Saffer   ISBN 0-937707-72-4

Vol. 54-The First Stromlo Symposium: The Physics of Active Galaxies
ed. Geoffrey V. Bicknell, Michael A. Dopita, and Peter J. Quinn   ISBN 0-937707-73-2

Vol. 55-Optical Astronomy from the Earth and Moon,
ed. Diane M. Pyper and Ronald J. Angione   ISBN 0-937707-74-0

Vol. 56-Interacting Binary Stars
ed. Allen W. Shafter   ISBN 0-937707-75-9

Vol. 57-Stellar and Circumstellar Astrophysics
ed. George Wallerstein and Alberto Noriega-Crespo   ISBN 0-937707-76-7

Vol. 58-The First Symposium on the Infrared Cirrus and Diffuse Interstellar Clouds
ed. Roc M. Cutri and William B. Latter   ISBN 0-937707-77-5

Vol. 59-Astronomy with Millimeter and Submillimeter Wave Interferometry
ed. M. Ishiguro and Wm. J. Welch   ISBN 0-937707-78-3

Vol. 60-The MK Process at 50 Years: A Powerful Tool for Astrophysical Insight
ed. C. J. Corbally, R. O. Gray, and R. F. Garrison   ISBN 0-937707-79-1

Vol. 61-Astronomical Data Analysis Software and Systems III
ed. Dennis R. Crabtree, R. J. Hanisch, and Jeannette Barnes   ISBN 0-937707-80-5

Vol. 62-The Nature and Evolutionary Status of Herbig Ae/Be Stars
ed. P. S. Thé, M. R. Pérez, and E. P. J. van den Heuvel   ISBN 0-937707-81-3

Vol. 63-Seventy-Five Years of Hirayama Asteroid Families:
The Role of Collisions in the Solar System History
ed. Yoshihide Kozai, Richard P. Binzel, and Tomohiro Hirayama   ISBN 0-937707-82-1

Vol. 64-Cool Stars, Stellar Systems, and the Sun, Eighth Cambridge Workshop
ed. Jean-Pierre Caillault   ISBN 0-937707-83-X

Vol. 65-Clouds, Cores, and Low Mass Stars
ed. Dan P. Clemens and Richard Barvainis   ISBN 0-937707-84-8

Vol. 66-Physics of the Gaseous and Stellar Disks of the Galaxy,
ed. Ivan R. King   ISBN 0-937707-85-6

Vol. 67-Unveiling Large-Scale Structures Behind the Milky Way,
ed. C. Balkowski and R. C. Kraan-Korteweg   ISBN 0-937707-86-4

Inquiries concerning these volumes should be directed to the:
 Astronomical Society of the Pacific
 CONFERENCE SERIES
 390 Ashton Avenue
 San Francisco, CA 94112-1722
 415-337-1100

ASTRONOMICAL SOCIETY OF THE PACIFIC
CONFERENCE SERIES

Volume 67

UNVEILING LARGE-SCALE STRUCTURES BEHIND THE MILKY WAY

Workshop at the Observatoire de Paris-Meudon
18-21 January 1994

Edited by
C. Balkowski and R. C. Kraan-Korteweg

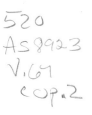

Table of Contents

List of Participants

ALTSCHULER D.	Arecibo Observatory, USA
BALKOWSKI C.	Observatoire de Paris, France
BARDELLI S.	Universita di Bologna, Italy
BOTTINELLI L.	Observatoire de Paris, France
BURTON B.	Leiden Observatory, Netherlands
CAPPI A.	Osservatorio di Bologna, Italy
CAYATTE V.	Observatoire de Paris, France
FABIAN A.	Institute of Astronomy, Cambridge, UK
FAIRALL A.	University of Cape Town, South Africa
FELENBOK P.	Observatoire de Paris, France
GIOVANARDI C.	Osser. Astrofisico di Arcetri, Firenze, Italy
GOUGUENHEIM L.	Observatoire de Paris, France
HAUSCHILDT-PURVES M.	Caltech Submill. Obs., Hilo, USA
HENDRY M.	University of Sussex, Brighton, UK
HENNING P.	Univ. New Mexico, Albuquerque, USA
HOFFMAN Y.	Racah Institute of Physics, Jerusalem, Israel
HUCHRA J.	Center for Astrophysics, Cambridge, USA
KERR F.	University of Maryland, College Park, USA
KRAAN-KORTEWEG R.	Kapteyn Lab., Groningen, Netherlands
LAHAV O.	Institute of Astronomy,Cambridge, UK
LEJEUNE T.	Observatoire de Strasbourg, France
LERCHER G.	Universitaet Innsbruck, Austria
LOAN A.	Institute of Astronomy, Cambridge, UK
LU N.	JPL/Caltech, Pasadena, USA
LYNDEN-BELL D.	Institute of Astronomy, Cambridge, UK
MAMON G.	I.A.P. and Observatoire de Paris, France
MARTIN J.-M.	Observatoire de Paris, France
MAUROGORDATO S.	Observatoire de Paris, France
MEURS E.	ESO, Garching bei Munchen, Germany
PANTOJA C.	Arecibo Observatory, USA
ROLA C.	Observatoire de Paris, France
SAITO M.	Kyoto University, Japan
SAUNDERS W.	Astrophysics, NAPL, Oxford, UK
SCHRODER A.	Astron. Inst., Univ. Basel, Switzerland
STASINSKA G.	Observatoire de Paris, France
STEWART R.	ATNF-CSIRO, Epping, Australia

TEUBER J.	Copenhagen Observatory, Denmark
VALENTIJN E.	Lab. Ruimteonderzoch, Groningen, Netherlands
VAN WOERDEN H.	Kapteyn Lab., Groningen, Netherlands
WAKAMATSU K.	Gifu University, Japan
WEINBERGER R.	Universitaet Innsbruck, Austria
WOUDT P.	Kapteyn Lab., Groningen, Netherlands
YAMADA T.	Kyoto University, Japan
ZUCCA E.	Ist. di Radioastron. del CNR, Bologna, Italy

Foreword

Many groups are currently looking in the Milky Way searching for galaxies on Schmidt plates, doing blind radio searches or identifying IRAS galaxies in order to **Unveil Large-Scale Structures behind the Milky Way.**

We thought the time was right to get these different groups together, to avoid duplication of searches in the same regions and to stimulate interactions. With our first goal we succeeded extremely well and we sincerely hope that our second aim will be fulfilled as well, i.e. that this workshop has lead to further and fruitful collaborations.

The meeting was divided in six different parts: General Introduction, Global Overviews, Optical Searches on Sky Surveys, Theoretical Presentations, Radio Surveys and IRAS Selected Samples. Donald Lynden-Bell concluded with comments on the results presented here. This was followed by a joint discussion.

Finally, we would like to elaborate a bit on the poster which was designed by Margie Walter and Tony Fairall in Cape Town. All of you must have seen the shepherd looking through the stars in many text books. It has been drawn by Camille Flammarion, a famous 19th-century French popularizer of Astronomy. One of his most popular books is 'l'Astronomie populaire'. He also published a lot of professional papers but his relationship with the 'official' astronomy was difficult. He was hired at Paris Observatory by Le Verrier in 1858, but was soon fired again because he published a popular book on 'La Pluralité des Mondes' in which he exposed his views on the real aim of astronomy: it should become astrophysics in order to merge the old astronomy with the cold mathematics. He returned to the Observatory only 14 years later. During that time he worked on double stars but he continued to publish popular books among which 'l'Astronomie des Dames'! His wish was to have his own observatory. He realized this in 1882 in Juvisy. Moreover, he founded the Journal 'l'Astronomie Populaire' and 'la Société Astronomique de France' for the public. This society still exists today.

C. Balkowski and R.C. Kraan-Korteweg

Acknowledgements

This meeting would not have been possible without the financial support of the Observatoire de Paris, the Kapteyn Laboratorium at Groningen, the CNRS and the University Paris 7.

We thank the secretaries of the department, N. Dreux, J. Plancy and E. Veia as well as F. Haslé and M. Clémino who helped during different phases of this meeting. Our special thanks go to S. Gordon for taking upon her the administrative and financial responsibilities of the meeting as well as for her help in preparing the Proceedings.

We are grateful to the Scientific Advisory Committee and to the Local Organising Committee.

Scientific Organising Committee

B. Burton	O. Lahav
A. Dekel	D. Lynden-Bell
A. Fairall	M. MacGillivray
L. Gouguenheim	M. Saitō
P. Henning	M. Strauss
F. Kerr	

Local Organising Committee

V. Cayatte

G. Mamon

J.-M. Martin

Part I

Introduction

Unveiling Large-Scale Structures Behind the Milky Way
ASP Conference Series, Vol. 67, 1994.
C. Balkowski and R. C. Kraan-Korteweg (eds.)

THE EARLY HISTORY OF GALAXY SEARCHES IN THE ZOA

FRANK J. KERR,
Astronomy Program, University of Maryland, College Park, MD 20742,
U.S.A.

ABSTRACT   From the earliest days of 21-cm work, the possibility of finding
something behind the Milky Way was considered by observers. This was
often expressed as 'the possibility of finding another M31'. We took a
number of 'stabs in the dark' at Parkes from time to time, and probably some
other observers did the same. Successful use of blind searching led to the
discovery of "high-velocity clouds" which were generally believed to be
related to the Galaxy, but nothing in the velocity range of spiral galaxies was
detected in this early work.
Eventually two significant developments increased the attractiveness of
looking for hidden galaxies at 21-cm. Firstly, receivers became much more
sensitive, bringing distant galaxies more clearly into their range. Secondly,
optical work has built up an extensive (but still partial) picture of the large-
scale structure of the extragalactic system, and this made more obvious the
desirability of finding galaxies in the hidden region to try to fill in the gaps in
the structural pattern.
We began our search for hidden galaxies in the summer of 1986, when we
showed that it was feasible (but very slow) to discover galaxies in this way.
Later observing sessions yielded more galaxies, and excited interest in moving
to a full-scale census of galaxies in the hidden region, preferably with
equipment which would enable a higher speed of searching.

I would like to start with a short personal note, which is based on my Southern
Hemisphere origin. The Australian Aborigines have a rich mythology concerning
objects in the sky, which they describe as their "Dreamtime", when their far-distant
ancestors went up to the heavens. There is however an important difference from
the analogous systems produced by Northern Hemisphere peoples, because they do
not have constellations. Instead, the prominent member of their sky is the Milky
Way, which is clearly the brightest object in the Southern Sky, as all travellers know.
My Australian upbringing has always pointed me towards the Milky Way as an
interesting object, firstly to study for its own sake and then later it was natural to
wonder what lay behind it, as the Milky Way interfered with extragalactic
investigations over about a quarter of the sky.
Many people have studied portions of the Zone of Avoidance to see whether
any galaxies could be found there by diligent searching. These studies have been
made in various wavelength ranges, but most work has been done in the optical

range. A detailed account of optical and infra-red searches was given by Renée
Kraan-Korteweg in a 1991 talk to the second DAEC Meeting, also at Meudon. The
motivation in almost every case was to obtain information which might be of use in
extending one or more large-scale structure features into the hidden region.

In these studies, the greatest impression was made by the discovery of Maffei
1 and 2. These were discovered at latitudes -0°.57 and -0°.33 from infra-red
photographs. Some people have described these as spirals, others as ellipticals. This
difference of opinion is a good example of the difficulty of observing in the ZOA.

Several thousand galaxies have now been recognized in the latitude range -9°
to +9°, mainly from studies of optical or infra-red photographs, but very few in the
innermost ±3° of latitude. It has often been thought that the data bank arising from
observations with the IRAS satellite would be a good place to pick out low-latitude
galaxies in the obscured region, because of the very large number of infra-red
objects listed in the catalog. However it is this large number which causes difficulty,
because the survey is confusion-limited in the low-latitude region, and individual
galaxies cannot be picked out.

From the earliest days of observations with the 21-cm hydrogen line in the
nineteen-fifties, it was recognized that this technique provided the possibility of
seeing galaxies behind the Milky Way. These considerations sometimes led to the
thought that perhaps we could find another M31. Our group at Parkes tried looking
at random points every now and again, but without any success in the early days,
largely because receivers at that stage weren't really sensitive enough.

The first positive results from blind searching at 21 cm in the ZOA came from
observations of high-velocity clouds, the name given to neutral hydrogen sources at
radial velocities exceeding about ±150 km s^{-1}. A number of these appear at, or pass
through, low galactic latitudes. There has been considerable controversy in the past
as to whether these are Galactic objects (such as clouds falling into the Galaxy), or
extragalactic objects nearby (such as the Magellanic Stream). There is probably
room for both types of object in the interpretation. The high velocity clouds have no
real connection with the higher-redshift objects of interest in the present context, but
are worthy of mention as a product of blind searching.

One unusual case was when Simonson in 1975 suggested that a widespread
object at intermediate velocities (50 to 120 km s^{-1}) was a very nearby galaxy.
However this interpretation has not been generally accepted, largely because no part
of the object could be seen in any other spectral range.

We began blind searching for hidden galaxies in 1986 in a series of
observations with the 300-foot telescope at Green Bank. We believe this was the
first program in which an extensive survey was set up with the express purpose of
determining whether it would be feasible to aim eventually to take a census of a
large number of galaxies (mainly spirals) in the hidden region. Two developments
made such a program interesting in 1986. One was the great improvement in the
sensitivity of receivers, which would increase the number of faint objects that could
be seen. The other was the great interest aroused in the large-scale structure of the
extragalactic universe, and the knowledge that many of these structural features
disappeared into the hidden region.

The plan for the survey was to follow lines of constant declination through the
ZOA, taking a new observation every one degree as the sky swept past the stationary
telescope. Profiles for the four most interesting objects in the first program are

shown in Figure 1. Later observations over succeeding years revealed a substantial number of new galaxies, showing that a 21-cm blind search was a good method for exploring the hidden region. Other observers have also made use of the blind search method, usually for specific parts of the ZOA.

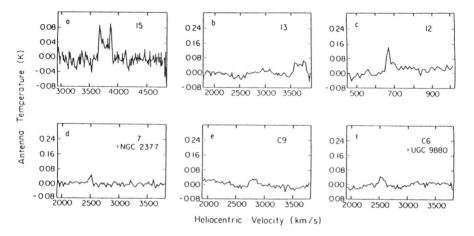

Figure 1. A selection of typical H$_I$ profiles. Figs. *a - d* are galaxies in the zone of avoidance: figs. *e* and *f* are galaxies n the clear region.

While the method is attractive, it has two principal shortcomings. One is that a very long time would be needed to explore the whole ZOA using the classical method of observing a single point in the sky at a time. Consequently a great deal of thought has been given to an antenna and receiver arrangement in which a substantial number of points can be observed at once, thereby greatly reducing the time required for a full survey.

The other shortcoming is the ease with which solar radiation and other interference can be picked up in the sidelobes of a normal antenna system. As a result the baseline tends to be variable, especially during the day, and the weakest sources cannot be clearly detected. The new Green Bank telescope, which has been specially designed to minimize the unwanted sidelobes, should be especially good for this type of low-level survey.

We can expect that large numbers of new galaxies will be discovered in the ZOA by the 21-cm blind-searching method.

Discussion

E. Meurs: Positive results with blind HI searches were obtained around 1986/ 1987. At that same time the first IRAS galaxy selections appeared. Was there any connection between these two observational efforts?

F. Kerr: The early HI results were compared with the IRAS listing, but the latter were too much limited by confusion in the lowest declinations.

Unveiling Large-Scale Structures Behind the Milky Way
ASP Conference Series, Vol. 67, 1994.
C. Balkowski and R. C. Kraan-Korteweg (eds.)

Introduction : Why and How to Probe the Zone of Avoidance ?

Ofer Lahav

Institute of Astronomy, Madingley Road, Cambridge CB3 0HA, UK

Abstract. The motivation and major ways for probing the Zone of Avoidance (ZOA) are reviewed. Galaxies hidden behind the ZOA may have important implications for the internal dynamics of the Local Group, for the origin of its motion relative to the Microwave Background, and for the connectivity of the large scale structure. Current direct ('observational') methods for exploring the ZOA include eye-balling of plates, source identification in the IRAS data base, and pointed and blind-search observations in 21 cm. Interesting regions identified so far include the two crossing points of the Supergalactic Plane by the Galactic Plane (at Galactic longitude $l \sim 135°$, near Perseus-Pisces, and $l \sim 315°$, near the Great Attractor), the Puppis cluster (at $l \sim 240°, cz \sim 1500$ km/sec) and the Ophiuchus cluster (at $l \sim 0°, cz \sim 8400$ km/sec). New promising wavelengths are the 2μ and the X-ray band. Indirect ('theoretical') approaches include 'Wiener reconstruction' from incomplete and noisy data, and using the peculiar velocity field as a probe of the mass distribution hidden behind the ZOA. The problem of source confusion at low Galactic latitude can be addressed by novel statistical methods, e.g. Artificial Neural Networks.

1. Introduction

Our own Galaxy is a major obstacle in probing the extragalactic sky. Local stars and dust block our view in most wavelengths, or cause confusion in identifying galaxies. If the 'Zone of Avoidance' (ZOA) covers Galactic latitudes $|b| < b_c$ then a solid angle $\omega = 4\pi \sin b_c$ is obscured. This corresponds typically to a fraction $\sim 20\%$ of the sky in the optical band and $\sim 10\%$ in the infrared.

As indicated by the term ZOA, the common view is that extragalactic studies should avoid this complicated region. Indeed, for some studies, e.g. for calculating correlation functions, it is not essential and even better to analyse regions of high Galactic latitude, as long as the sample is a 'fair sample'. However, for understanding other issues of the large scale structure it is crucial to unveil the whole-sky galaxy distribution. An earlier review (Kraan-Korteweg 1993) summarized previous studies, and the history of the subject is reviewed by Kerr in this volume. Here we shall attempt to summarize present and future methods for exploring the ZOA, and to show that the ZOA is not actually such an 'obscured' topic. In this meeting, we mainly refer to the ZOA in the context of the study of galaxy distributions. More generally, the ZOA also affects other

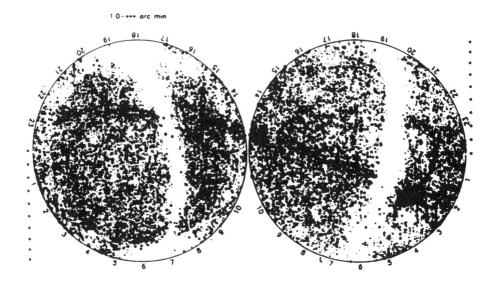

Figure 1.   An equal area projection of the 2 hemispheres in Equatorial
coordinates of galaxies larger than 1 arcmin from the UGC, ESO and
MCG catalogues. The most prominent filament across the plot is the
Supergalactic Plane. The empty bands are due to obscuration by the
Milky way (after Lynden-Bell & Lahav 1988).

extragalactic measurements, e.g. the Cosmic Microwave Background and the
X-ray background.

2.   Implications of hidden matter behind the Milky Way

2.1.   The connectivity of structure

Figure 1 shows the two hemispheres (in Equatorial coordinates) plotted in equal-
area projection. The galaxies shown are from the UGC, ESO and MCG catalogues. The two empty bands are the ZOA. The Supergalactic Plane is seen in its
full glory, but its full extent is hidden by the ZOA. There are also several other
filaments running perpendicular to the ZOA (e.g. in the Southern hemisphere
near Hydra, dubbed together as the 'dinosaur's foot' by Donald Lynden-Bell).
This figure indicates that if we wish to know the extent of the Supergalactic
Plane (and to learn if it is a 'plane' at all) we do need to unveil the structure
behind the ZOA.

    As pointed out by Kraan-Korteweg (1993), 5 out of the 8 apparent brightest
galaxies lie in the ZOA : CenA, IC342, Maffei I and II and N4945. Since galaxies

are clustered, it would not be too surprising to discover more galaxies in the neighbourhood of these known galaxies.

2.2. The internal dynamics of the Local Group

It is intriguing to speculate on the existence of a nearby $M31$-like galaxy behind ZOA. Such a nearby galaxy, or several galaxies, would change dramatically our understanding of the dynamics of the Local Group (LG), currently believed to be dominated by the Milky Way and Andromeda. Discovering new members of the Local Group would also affect the deduction of the mass and the age of the Local Group from its dynamics (Kahn & Woltjer 1959; Lynden-Bell 1982) and the reconstruction of LG galaxy motions (Peebles 1990).

2.3. The origin of motion of the Local Group

Understanding the origin of motion of the Local Group relative to the Cosmic Microwave Background (CMB) radiation is of major importance for verifying the gravitational instability picture and for the deduction of the density parameter Ω_0 and 'biasing' of the galaxies relative to the underlying mass distribution.

The CMB dipole is very accurately measured and is commonly interpreted as being due to the motion of the sun relative to the CMB radiation (at 370 km/sec towards $l = 264°, b = 48°$; Kogut et al. 1993). This motion is converted to the motion of the Local Group relative to the CMB (at about 600 km/sec towards $l = 268°, b = 27°$) by taking into account the rotation of the sun around the Galaxy and the relative motions of the Galaxy and Andromeda (see e.g. Yahil, Tammann & Sandage 1977; Lynden-Bell & Lahav 1988). Even if no nearby $M31$-like galaxy is to be found behind the ZOA, and the above calculation holds, then one needs to understand the sources of the motion of the Local Group at 600 km/sec. Some of the sources are probably behind the ZOA.

The expected peculiar velocity is calculated, assuming linear theory (Peebles 1980) by summing up the contribution from masses M_i represented by the galaxies

$$\mathbf{v} \propto \frac{\Omega_0^{0.6}}{b} \sum_i \frac{M_i}{r_i^2} \hat{\mathbf{r}}_i \, , \tag{1}$$

where Ω_0 is the density parameter and b is the bias parameter (reflecting that light may not be a perfect tracer of the mass). Commonly the masses are assumed to be either equal or proportional to the galaxy luminosities (hence only the observed flux is required). The dipole in the distribution of IRAS and optical galaxies lies within $10° - 20°$ of the CMB dipole (e.g. Lynden-Bell, Lahav & Burstein 1989, Strauss et al . 1992), but the convergence of the dipole with distance is still an open question. In part this uncertainty is due to shot-noise, the finite depth of the samples, and redshift distortion, but another factor may well be the missing data behind the ZOA.

It is curious that the two major superclusters on opposite directions on the sky, the Great Attractor and Perseus-Pisces, which are playing the role in the tug of war' on the Local Group, are near the ZOA. Moreover, two major galaxies, CenA and IC342, are also in these opposite directions near the ZOA. Another example of a possible contribution to the Local Group motion is due the Puppis cluster at $l \sim 240°$; $b \sim 0°$; $cz \approx 1500$ km/sec. By comparison with the 2Jy

IRAS survey, it has been suggested (Lahav et al. 1993) that Puppis (which lies below the Supergalactic Plane) may contribute at least 30 km/sec to the motion of the Local Group *perpendicular* to the Supergalactic Plane ($V_{SGZ} \approx -370$ km/sec). Together with the Local Void (above the Supergalactic Plane, also crossed by the ZOA) and Fornax and Eridanus (below it) this may explain of the origin of the so-called "Local Anomaly".

3. Direct 'Observational' methods

3.1. Understanding the Milky Way and Extinction

The first step in exploring the extragalactic sky behind the ZOA is to understand our own Galaxy. The extinction and confusion with stars become more and more difficult as one approaches the Galactic Plane. Even the detailed extinction map of Burstein & Heiles (1982) only covers $|b| > 10^\circ$. New studies of the Galactic HI distribution (see Burton in this volume) are most useful to predict extinction, in particular if combined with the IRAS 100 μ emission maps and other Galactic measurements. The next step is to understand how Galactic extinction affects galaxy magnitudes and diameters. Cameron (1990) showed how to obscure high-latitude galaxies artificially, an approach which needs to be further developed.

3.2. Eye-balling of plates and optical redshift surveys

Earlier searches for galaxies on plates were carried out by Böhm-Vitense (1956) and Weinberger (1980). Focardi et al. (1984) suggested that Perseus-Pisces is connected to A569, as was later supported by radio and optical redshift measurements (Hauschildt 1987, Chamaraux et al. 1990). The region of Puppis was searched by Saito et al. (1990, 1991) on infrared film copies. Kraan-Korteweg (1993; also in this volume) carried out a systematic surveys on IIIaJ film copies of the ESO/SRC-survey in the directions of Hydra, and Hau surveyed Palomar red plates in the region centred on $l \sim 135^\circ$, where the Supergalactic plane is crossed by the GP.

The above are only some examples of many elaborate surveys, revealing thousands of new galaxies. They also provide target lists for follow-up of optical and radio redshift measurements (e.g. Fairall and Huchra in this volume), which then reveal the 3-dimensional structure behind the ZOA. However, the different searches are far from being on equal footing, and their interpretation is subject to selection effects, e.g. due to the plate material used, and the poorly known extinction at low latitudes.

The elaborate procedure of eye-balling calls for new automatic methods which will allow separating stars from galaxies in deep samples such as the APM (Maddox et al. 1990) and the 2μm survey (Mamon in this volume). Another approach (Odewahn et al. 1991) utilizes Artificial Neural Networks which are 'trained' to separate galaxies from stars in a non-linear way (the method can also be used to classify galaxies into morphological types; Storrie-Lombardi et al. 1992). Such methods allow reproducibility and objective classification.

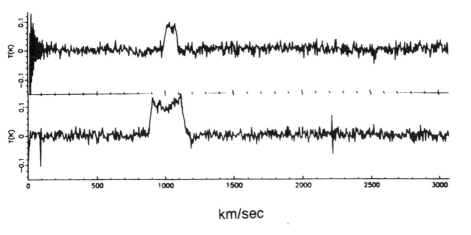

km/sec

Figure 2.   HI spectra from the Dwingeloo survey. The galaxy spectrum at the bottom is of NGC3359 at high Galactic latitude ($l =$
$144°, b = 49°$), while the spectrum at the top is of a galaxy behind the
ZOA ($l = 95°, b = 2°$; number 34 in Henning 1992). (Burton, Ferguson,
Hau, Henning, Kraan-Korteweg, Lahav, Loan & Lynden-Bell; work in
progress.)

3.3.   21 cm

The search for galaxies behind the GP in the 21 cm line of neutral hydrogen (HI),
was pioneered by Kerr & Henning (1987). They carried out (using the Green
Bank 91m radio-telescope) blind observations towards 1900 points and detected
16 new galaxies. The spectra of the galaxies detected did look similar to those of
galaxies at high Galactic latitude, indicating that the ZOA is almost transparent
in HI measurements. Some show the typical spectral 'double horned' signature
of a spiral galaxy. Others may well represent other populations, e.g. of dwarf
galaxies and low surface brightness galaxies. This work has been extended by
Henning (1992; and in this volume) and by other groups.

A new survey began recently using the Dwingeloo 25-m telescope in the
Netherlands. This radio-telescope, recently used to map the Milky Way in HI
(see Burton in this volume) is now dedicated to do blind search and pointed
observations of galaxies out to 4000 km/sec as a collaboration of Cambridge,
Dwingeloo, Groningen and Leiden. Figure 2 shows examples of galaxy spectra
from this survey at high and low Galactic latitude, illustrating that galaxies can
be detected behind the Galactic plane. A deeper new multi-beam whole-sky
survey is planned using the Parkes 64-m and the Lovell 76-m (see Stewart in
this volume.)

An alternative to the blind search approach is to have pointed 21 cm observations towards galaxy candidates selected from the IRAS Point Source Catalog
or from optical plates. The two approaches are complementary. The blind search
is systematic but rather slow, and includes population of dwarf and low surface
brightness galaxies. The IRAS-selected HI search is biased towards spiral galax-

ies, and is limited at very low Galaxy latitudes where IRAS is confusion limited. The selection effects in HI searches are the cutoff in redshift, the velocity resolution, and problems of solar interference.

3.4. IRAS

The IRAS Point-Source Catalog (PSC) has been exploited in recent years to provide galaxy candidates behind the ZOA (e.g. Lu et al. 1990, Yamada et al. 1993, and others in this volume). Different authors proposed different colour selection criteria to pick up galaxies from the PSC. These candidates were then followed up by HI radio surveys or by inspection of plates. The advantage of this approach is in producing a uniform sample, which can be related to the rest of the sky. Possible problems are confusion with Galactic sources and bias towards spiral galaxies.

3.5.   2 μm

This new wavelength for extragalactic surveys is promising for two reasons: (i) it probes the old 'stable' stellar population in galaxies and (ii) it little suffers from Galactic extinction. On the other hand, it suffers confusion with Galactic sources at low Galactic latitude.

Two major surveys are now carried out in 2μm. The 2 Micron All-Sky Survey (2MASS), described by Huchra in this volume, which will cover the sky at 3 bands. This survey will yield a sample of about 100000 galaxies down to 13-14 K magnitudes within $10°$ of the GP. The second survey, DENIS, described in this volume by Mamon, will cover the Southern sky.

3.6.   X-ray

This wavelength is discussed by Fabian in these Proceedings. X-ray selected clusters are more easily detected than optical clusters at low Galactic latitude (e.g. Lahav et al. 1989). The whole-sky HEAO1 and Rosat surveys are in particular useful for the ZOA problem.

4.   Indirect 'theoretical' methods

4.1.   Wiener reconstruction of all sky surveys

Previous corrections for the unobserved ZOA (e.g. for the dipole calculation) were done, in a somewhat ad-hoc fashion, by populating the ZOA uniformly according to the mean density, or by 'shifting' or interpolating the structure below and above the Galactic Plane to create 'mock sky' in the ZOA (e.g. Lynden-Bell et al.1989, Strauss et al. 1992, Hudson 1993). An alternative approach allows mask inversion by regularization with a 'Wiener filter' (the ratio of signal to signal+noise), in the framework of Bayesian statistics and Gaussian random fields, by assuming a prior model for the power-spectrum (or correlation function) of the galaxies. Lahav et al. (1994) have applied this method, utilizing spherical harmonics, to the projected IRAS 1.2Jy survey. Their whole-sky noise-free reconstruction confirms the connectivity of the Supergalactic plane across the ZOA (at $l \sim 135°$ and $l \sim 315°$) and the Puppis cluster at $l \sim 240°$. Detailed discussion of this method and pictorial reconstructions appear elsewhere in this volume (see contributions by Hoffman and Lahav).

4.2. Dynamical reconstruction of the mass behind the ZOA

Instead of looking at the distribution of galaxies at the ZOA, one can use the peculiar velocity field on both sides the ZOA to predict the *mass*-density distribution behind the Galactic Plan. The Potent method (Dekel, Bertschinger & Faber 1990) allows to recover the potential from line-of-sight peculiar velocities and hence the mass-density field. Kolatt, Dekel & Lahav (1994) have used the method to look in the direction of the ZOA.

The main dynamical features found at a distance $r \sim 4000$ km/sec are (a) the peak of the Great Attractor connecting Centaurus and Pavo at $l \sim 330°$, (b) a moderate bridge connecting Perseus-Pisces and Cepheus at $l \sim 140°$, and (c) an extension of a large void from the southern Galactic hemisphere into the ZOA near the direction of Puppis, $l \sim 220° - 270°$. They found a strong correlation between the mass density and the IRAS and optical galaxy density at $b = \pm 20°$, which indicates that the main dynamical features in the ZOA should also be seen in galaxy surveys through the Galactic Plane. Note that at 4000 km/sec the Potent resolution of 1200 km/sec corresponds to 17°, so one can only reconstruct the gross features across the ZOA, rather than the individual clusters. The gravitational acceleration at the Local Group, based on the *mass* distribution out to ~ 6000 km/sec, is found to be strongly affected by the mass distribution in the ZOA: its direction changes by 31° when the $|b| < 20°$ ZOA is included, bringing it to within $4° \pm 19°$ of the CMB dipole.

5. New individual structures behind the ZOA

Various structures behind the GP are discussed by different authors in detail in this volume. Here we only point out some examples of interesting regions.

5.1. The Great Attractor and A3627

From the study of the peculiar velocity field of elliptical galaxies Lynden-Bell et al. (1988) predicted, using a simple spherical model, that the centre of the 'Great Attractor' is at $l = 307°, b = 9°$, i.e. behind the ZOA. It is interesting that several recent studies also suggest similar directions. Kraan-Korteweg & Woudt (these Proceedings) have studied this region by visual inspection of IIIaJ copies of the ESO/SRC sky surveys and redshift measurements. They find a peak centred on the rich ACO cluster A3627 ($l = 325°, b = -7°, cz = 4300$ km/sec). Reconstruction methods, although with smoothing larger than cluster-scale, also suggest an enhancement in this neighbourhood. Kolatt et al. (1994) find a peak in *mass-density* in the Potent reconstruction at ($l = 320°, b = 0°, cz = 4000$ km/sec) and Hoffman (this volume) also finds a peak in the galaxy distribution in that region.

5.2. Puppis

This cluster was recognized independently by Kraan-Korteweg & Huchtmeier (1992), Yamada et al. (1992), and Scharf et al. (1992). To quantify the impor-

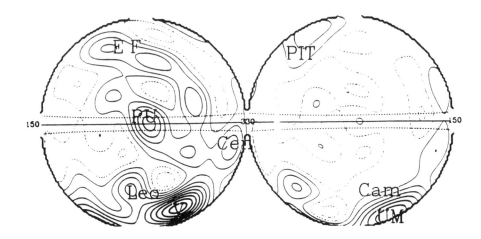

Figure 3.   Spherical Harmonic reconstruction with coefficients up to
$l_{max} = 10$, of galaxies with 500 km/sec $< cz_{LG} < 3000$ km/sec in
the 2Jy IRAS ($|b| > 5°$) combined with our Puppis sample ($|b| < 5°$).
Plots are equal area hemispheres with the left-hand side plot centred
on Galactic $l = 240°$, $b = 0°$. The Galactic Plane runs horizontally
across the plots (solid line), dashed lines bound the region $|b| < 5°$ and
longitudes are indicated. The South Galactic hemisphere is at the top.
The lightest solid contour is at the mean, dashed and solid contours
indicate densities below and above the mean respectively. Contour separation is 3 times the shot-noise level. Associations with local structures are labelled: V - Virgo, PU - Puppis, F - Fornax E - Eridanus,
Leo - Leo, Cen - Centaurus, PIT - Pavo-Indus-Telescopium, Cam -
Camelopardalis and UM - Ursa Major. (From Lahav et al. 1993).

tance of Puppis relative to other structures in the local universe Lahav et al.
(1993) supplemented the IRAS 2 Jy $|b| > 5°$ redshift survey (Strauss et al. 1992)
with an IRAS-selected sample in the direction of Puppis ($|b| < 5°; 230° < l <
260°$), which consists of 32 identified galaxies, 12 of them with measured red-
shift. It was found that the projected number counts of galaxies brighter than 2
Jy in Puppis is about half that of Virgo. A Spherical Harmonic reconstruction
(Figure 3) shows that out to a distance of 3000 km/sec Puppis is second only
to Virgo. However, Puppis does not seem to be a virialized cluster, and it is
not seen in the HEAO1 and ROSAT X-ray maps (K. Jahoda and H. Böhringer,
private communication)

5.3.   Ophiuchus

This cluster, behind the Galactic Centre (!), has been studied by Wakamatsu
(this volume, and references therein). His catalogue of 6000 galaxies suggests

that there is a big supercluster around the Ophiuchus cluster ($l = 0°, b = 8°, cz \approx$ 8500 km/sec). This is also one of the brightest known X-ray clusters (e.g. Wakamatsu & Malkan 1980, Lahav et al. 1989 and Fabian in this volume). Moreover, the Ophiuchus supercluster seems to be connected to the Hercules supercluster. Djorgovski et al. (1990) speculated that Ophiuchus is also connected to the Sagittarius cluster (at $l = 359°, b = 8°, cz \approx 8600$ km/sec).

6. Discussion

This is the first meeting dedicated to the ZOA problem, bringing together researchers who are using different methods and working at different wavelengths. The following are some of the important issues calling for discussion:

- How to produce a combined Galactic extinction map based on HI, IRAS etc. ?

- Who is observing what and where in the ZOA ?

- What are the selection effects in optical, IRAS, 2μm, 21 cm and X-ray surveys ?

- How to combine different surveys to produce a uniformly sampled map ?

- How to implement algorithms for automatic separation of galaxies from stars at low Galactic latitudes ?

Acknowledgments. I am grateful to H. Ferguson, R. Kraan-Korteweg, A. Loan and D. Lynden-Bell for their comments on this manuscript, and to my collaborators to some of the work described here for their contribution and helpful discussions.

References

Böhm-Vitense, E. 1956, PASP, 68, 430

Burstein, D. & Heiles, C. 1982, AJ, 87, 1165

Cameron, L.M. 1990, A&A, 233, 16

Chamaraux, P., Cayatte, V., Balkowski, C., & Fontanelli, P. 1990, A&A, 229, 340

Dekel, A., Bertschinger, E. & Faber, S.M. 1990, ApJ, 364, 349

Djorgovski, S., Thompson, D.J., De Carvalho, and Mould, J.R. 1990, AJ, 100, 599

Focardi, P., Marano, B. & Vettolani, G., 1984, A&A, 136, 178

Hauschildt, M., 1987, A&A, 184, 43

Henning, P.A. 1992, ApJS, 78, 365

Hudson, M. 1993, MNRAS, 265, 72

Kahn, F.D. & Woltjer, L. 1959, ApJ, 130, 705

Kerr, F. J., and Henning P.A. 1987, ApJ, 320, L99

Kogut, A. et al. 1993, ApJ, 419, 1

Kolatt, T., Dekel, A. and Lahav, O. 1994, MNRAS, submitted

Kraan-Korteweg, R. C. 1993, in *The Second DAEC meeting on the Distribution of Matter in the Universe*, ed. G. Mamon & D. Gerbal (Meudon), p. 202.

Kraan-Korteweg, R. C. & Huchtmeier, W.K., 1992, A&A, 266, 150

Lahav, O. Edge, A.C., Fabian, A.C. & Putney, A. 1989, MNRAS, 238, 881

Lahav, O., Yamada, T., Scharf, C., A. & Kraan-Korteweg, R.C. 1993, MNRAS, 262, 711

Lahav, O., Fisher, K.B., Hoffman, Y., Scharf, C.A. & Zaroubi, S. 1994, ApJ, 423, L93

Lu, N.Y., Dow, M.W., Houck, J.R., Salpeter, E.E. & Lewis, B.M. 1990, ApJ, 357, 388

Lynden-Bell, D. 1982, Observatory, 102, 7

Lynden-Bell, D., Faber, S.M., Burstein, D., Davies, R.L., Dressler, A., Terlevich, R.J. & Wegner, G. 1988, ApJ, 326, 19

Lynden-Bell, D. & Lahav, O. 1988, in *Large Scale Motions in the Universe, Proceedings of the Vatican Study Week*, ed. G. Coyne & V.C. Rubin, Princeton University Press, Princeton

Lynden-Bell, D., Lahav, O. & Burstein, D. 1989, MNRAS, 241, 325

Maddox, S.J., Sutherland, W.J., Efstathiou, G. & Loveday, J. 1990, MNRAS, 243, 692

Odewahn, S.C., Stockwell, E.B., Pennington, R.L., Humphreys, R.M., Zumach, W.A. 1991, AJ, 103, 318

Peebles, P.J.E. 1980, *The Large Scale Structure of the Universe*, Princeton University Press

Peebles, P.J.E. 1990, ApJ, 362, 1

Saito, M. et al. 1990, PASJ, 42, 603

Saito, M. et al. 1991, PASJ, 43, 449

Scharf, C., Hoffman, Y., Lahav, O., & Lynden-Bell, D. 1992, MNRAS, 256, 229

Storrie-Lombardi, M., Lahav, O., Sodré & Storrie-Lombardi, L.J. 1992, MNRAS, 259, 8p

Strauss, M.A., Yahil, A., Davis, M., Huchra, J.P., & Fisher, K. 1992, ApJ, 397, 395

Wakamatsu, K. & Malkan, M. 1981, PASJ, 33, 57

Weinberger, R. 1980, A&AS, 40, 123

Yahil, A., Tammann, G.A. & Sandage A. 1977, ApJ, 217, 903

Yamada, T., Takata, T., Djamaluddin, T., Tomita, A., Aoki. K., Takeda, A. & Saito, M., 1993, MNRAS, 262, 79

Discussion

T. Yamada: You said that one of the motives of investigating the galaxy distribution behind the ZOA is to construct a '3-D fair sample'. What do you mean by 'fair sample' in this case? If you see a deeper universe (as deep as you need!), you will get a 'fair sample' at high latitude and you might not have to study the structure behind the Milky Way?

O. Lahav: The definition of a 'fair sample' depends of course on the statistic of interest. If we want to study nearby filaments, the structure behind the ZOA is important. Furthermore if we go very deep we may encounter new problems, e.g. evolution.

C. Balkowski: Could you tell more about the new method for star/galaxy separation?

O. Lahav: Computer algorithms called Artificial Neural Networks 'learn' (essentially by least square minimization with respect to free parameters called 'weights') to classify images into stars and galaxies from a set of examples for which the answer is already known (e.g. from a human expert). After a network has been 'trained' (by fixing the weights), new objects (previously unclassified) are presented to the network. The network provides probabilities for an object being a star or a galaxy.

L. Gouguenheim: I have an additional remark about selection effects - the majority of neutral hydrogen studies starting from IRAS sources, selected from their IR colours, one still focused on optically identified objects, because of the better detection rate. Galaxies are then found in region of low obscuration.

O. Lahav: Blind searches in HI may overcome some of these problems.

Part II

Global Overviews

Unveiling Large-Scale Structures Behind the Milky Way
ASP Conference Series, Vol. 67, 1994.
C. Balkowski and R. C. Kraan-Korteweg (eds.)

Visualization of Nearby Large-Scale Structures

A.P. Fairall, W.R. Paverd and R.P. Ashley

Departments of Astronomy and Computer Science, University of Cape Town, Rondebosch, 7700 South Africa

Abstract. A graphics workstation has been used to display the distribution of galaxies, extracted from the first author's "Southern Redshift Catalogue", in redshift space. At the conference, viewgraph transparencies, slides, and video sequences were presented. Emphasis has been on nearby ($cz < 20000$ km s^{-1}) structures, identified by colour coding, seen adjacent to the Southern Milky Way. The most dominant structures are two "great walls" - the Sculptor and Fornax Walls that run roughly parallel to each other. The Sculptor Wall has a thickness of 1000 km s^{-1}, within which it exhibits a cellular structure. If the nearer Fornax Wall has similar structure, then one can probably see the Centaurus supercluster as a continuation of this structure on the other side of the Milky Way. Thus our galaxy, and the nearby Fornax, Hydra I and Centaurus clusters, may be embedded within a "Centaurus Great Wall".

1. Introduction

Present-day mapping of large-scale structures reveals a tendency for many galaxies to conglomerate within "great wall" structures. The original "Great Wall", in the northern skies, was identified from the Center for Astrophysics survey (Geller and Huchra 1989); similarly, da Costa (1992) called attention to the "Great Southern Wall". The intriguing results of Broadhurst et al (1990) may suggest the concentration of galaxies in such walls at uniform intervals of redshifts.

In this paper, we shall attempt to review what is generally known of nearby ($cz < 20000$ km s^{-1}) structures, including great walls, with particular emphasis towards the southern sky. A complementary paper, that concerns the northern sky, is given, at this conference, by Marzke and Huchra.

As its title suggests, our conference presentation centered on visual materials - chiefly colour slides (high resolution) and video sequences (rotations, fly-throughs etc). Unfortunately, it is not possible to include such material with the conference proceedings. The accompanying figures (used as introductory viewgraphs in the presentation) will serve as adequate reference, though they cannot, of course, demonstrate the findings reported below. In any case, the material can best be seen on the high-resolution computer monitor - for which the slides and video were only a poorer substitute.

Figure 1 gives a schematic impression of nearby structures. At the top of the diagram, the well-known Great Wall, with the Coma Cluster embedded in it,

Figure 1.   A schematic diagram to indicate the approximate distribution of galaxies in the Supergalactic plane (for cz <7500 km s^{-1}). The upper half of the diagram is the Northern Galactic Hemisphere, the lower half the Southern Galactic Hemisphere. These two portions are divided by the obscuring band of the Milky Way. See text for discussion.

is seen in cross section. Although this structure connects to other neighbouring structures, the wall is orientated so as not to cross the Galactic Plane (GP). On the right hand side of the diagram is the Perseus-Pisces Supercluster, which is cut by the Northern GP. Nevertheless, the bulk of it lies in the Southern Galactic Hemisphere (SGH). It is the view of one of us (Fairall 1993) that this too is a "great wall" structure, but the wall is seen nearly edge-on in the sky. However, our chief interest here lies with Centaurus supercluster, and its extensions towards Hydra and Pavo, which lie on opposite sides of the GP.

Our own local (Virgo) supercluster is really a further extension of Centaurus, as there is a continuous distribution of galaxies connecting it to Centaurus. However, as the diagram suggests, a major portion of this structure is apparently obscured by the Southern Milky Way. Finally, though the diagram is not suitably orientated, it should be noted that the Great Wall, in the Northern Galactic Hemisphere (NGH), is roughly at right-angles to the Sculptor "Great Wall" (mainly SGH). Results reported below also highlight the tendency for right-angled intersections and parallel structures to occur.

Figure 2 illustrates important structures in the southern skies, in relation to the obscuration of the Milky Way. Here, as in the remainder of the paper, the data is drawn from the first author's "Southern Redshift Catalogue"(hereafter SRC - see other paper by Fairall, with references, this conference) for galaxies south of Decl = 0°.

It is important to realise that the data is that extracted from the literature - and is not necessarily complete to a magnitude or diameter limit (see also Discussion). Thus concentrations of galaxies that run in radial fashion (eg that labelled "Continuation of Sculptor Wall?" in Figure 2) are suspect, since they may be artificially created by an intensive study in a favoured small region in the sky. However, features that run across the line of sight have to be real, since they involve an extended region of sky for which the redshifts could not have been pre-selected. In the absence of the colour slides and video, Figure 2 will serve for reference below.

2. Graphical display

The SRC data was converted to rectangular coordinates and is displayed using custom-written software. The software was written using the Inventor three-dimensional graphics library, and runs on a *Silicon Graphics Indigo 2* workstation with *Extreme* graphics hardware.

The data is placed in a three-dimensional redshift environment that allows the user to interact with it. The interaction includes the ability to do real-time rotations, to fly through the data, and to display slices of arbitrary thickness and orientation. Each galaxy is represented by a single pixel on a 1280x1024 display. Previously recognized features are colour coded to make for better identification. These features can also be enclosed in 3D polyhedra to make identification easier. The features can viewed in the context of the entire database, or with the all galaxies that lie outside recognized features removed.

Each void is labeled with a three-dimensional name situated at the center of the void. The names are visually correct in terms of perspective, in other words, voids that lie further away can be identified by their smaller labels.

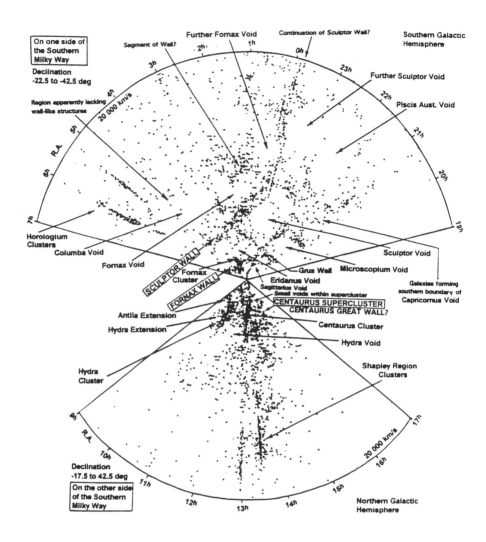

Figure 2.   The distribution in redshift space (to $cz = 20000$ km s^{-1}) of galaxies from the Southern Redshift Catalogue. Thick Declination "slices" show data on either side of the southern Milky Way. Almost all of the established features are identified; those that do not appear in the figure are the Pavo extension of the Centaurus supercluster (a great concentration of galaxies that forms the southern boundary of the Microscopium Void), the Octans Void (R.A. 4h, Decl. -80°, cz 3000 km s^{-1}) and the Apus Void (4h, -80°, 8000 km s^{-1}). This diagram may be used for reference to most of the features described in the text.

The graphics workstation provides an excellent tool for visually examining the large-scale distribution of galaxies. In particular, it allows the data to be viewed from any angle, as readily as if it were frozen within a transparent cube that could be held in the hand. One can therefore search for subtle features, while small rocking rotations emphasize the three-dimensional aspects of the data.

3. Results

Many large-scale structures and voids in the southern sky have been reported (since 1983) by the first author, using plots (slices) of SRC data. A parallel effort has been carried out with the statistically controlled data of the Southern Sky Redshift Survey (da Costa et al 1988, Pellegrini et al 1990). Identified features are indicated in Figure 2.

3.1. The Sculptor Wall

The largest, most clearly defined structure in the display, is the "Sculptor" or "Great Southern Wall". This structure is at least 8000 km s^{-1} long (in redshift space) - and possibly even as long as 20000 km s^{-1}, with a width of 5000 km s^{-1} (Fairall 1993). More important, perhaps, is that it is about 1000 km s^{-1} thick. The galaxies that form the wall are not distributed uniformly; rather, when viewed with a slight rocking rotation, a cellular structure within the wall, seemed apparent. In other words, there seemed to be voids, with diameter of several hundred km s^{-1}, embedded within the wall - an order of magnitude smaller than the large voids or underdense regions that separate the great walls themselves. Similar "holes" are apparent in the northern "Great Wall" (Ramella et al 1992 - as confirmed during this presentation by Huchra).

3.2. The Fornax Wall

A second wall lies at lower redshift - we have labelled it here as the Fornax Wall, since the Fornax Cluster lies within it. It is remarkable that this second wall runs virtually parallel ($< 5°$) to the Sculptor Wall - as can be seen in Figure 2, and confirmed on the graphical display. The planes of the walls are however tilted by 30±5 ° relative to one another. As with the Sculptor Wall, the member galaxies do not show a uniform distribution - but what is seen of the Fornax Wall in the SGH is somewhat thinner than the thick Sculptor Wall. However, the Wall is positioned up against the Southern Milky Way. One of the video sequences shown was a trip along the galactic plane (at constant cz =3000 km s^{-1}) - this gave the impression of the wall "scraping against" the band of obscuration of the Milky Way.

3.3. The Fornax Wall and the Centaurus Supercluster as a single entity

The finding most relevent to the topic of this conference is as follows: If the wedge of obscuration of the Milky Way is ignored, then the Fornax Wall (in the SGH) and the Centaurus supercluster (in the NGH) appear to be one and the same entity. Furthermore, whilst the geometric form of the Centaurus supercluster can at best be described as a sort of pancake, the combined form more closely resembles a great wall, that, like the Sculptor Wall, has considerable cellular texture embedded within it. Thus, it could be that our galaxy itself lies within a great wall - for which one might tentatively propose the name "Centaurus Great Wall".

If so, the GP cuts a cross section of this wall. Given the internal cellular structure, the distribution within this cross section would be far from uniform - indeed the strongest concentrations are in Hydra-Antlia and Centaurus (see Woudt, this conference). Furthermore, such internal cellular structure would explain the small voids within the Centaurus-Hydra concentrations (eg see Figure 2).

3.4. Other Walls

In view of the existence of the two great parallel walls in the SGH, a search was made to see if a third one was not present. There is a suggestion - a short segment at R.A. 1.5h, Decl -35°, cz 10000 km s^{-1} - of such a third parallel wall, but detailed examination failed to reveal any continuation. This is curious because the density of observations in this general direction should have detected such a structure (as also noted by da Costa - private communication). Perhaps it is relevent that the rich Horologium clusters are not far off - almost as though their presence disrupted the surrounding region. Yet, if so, this would contradict other situations where clusters (eg Coma and Fornax) are embedded in walls.

The display did confirm the presence of other walls that run more or less at right angles (90° ± 10°) to the two parallel walls. The most obvious is the Grus Wall (Maurellis et al 1990a).

3.5. The Eridanus Void

"Fly through" manoevers performed on the graphical display were used to examine voids and to search for previously unrecognised interconnections. The previous "Octans Void" (SRC 1988) was found to be a continuation of the very nearby (cz =2500 km s^{-1}) Eridanus Void (see Figure 2).

This is interesting since it makes the Eridanus Void a long tubular void that starts in the northern skies as the void in front of the Perseus-Pisces supercluster, then runs south (and is seen in cross section in Figure 2) and heads off in the direction of the South Celestial Pole - previous Octans Void (at cz =3000 km s^{-1}) which probable interconnects with the more distant Apus Void (cz =8000 km s^{-1}) in the same direction. The naming of this void after the long and winding constellation Eridanus is therefore appropriate. This void is not particularly clean; there are a spinking of galaxies throughout and at times the sides seem to "squeeze and constrict" the tube, but there are apparently no walls of galaxies to subdivide its length.

3.6. Other Voids

The Sculptor and Microscopium Voids remain the "cleanest" in the southern sky, their central cores being apparently totally devoid of galaxies (see Figure 2) outside of which there is a sprinking of inlying galaxies from the surrounding walls.

A new void was identified at R.A. 0h20m, Decl -30°, cz 12000 km s^{-1}. It seems appropriate to label it as the "Further Fornax Void" - see Figure 2.

4. Concluding Remarks

Clearly the flexibility in the way the data can be viewed, on the graphics workstation, is a significant improvement over fixed "slices" in R.A. or Declination (or galactic coordinates). It may permit the identification of subtle features and finer three-dimensional structures. The discovery of small-scale cellular structure (ie not just holes) within southern "great walls" - as described in the previous section - seems to be one of the most important results from this investigation. The finding is, of course, tentative. More data is needed, but this project is an ongoing process that accompanies the update of the SRC.

The graphical display is also ideal for tracing the continuity of features - such as the extension of the Eridanus Void. Also, and most relevent to this conference, the suggestion, made above, that the Fornax Wall (in the SGH) and the Centaurus supercluster (in the NGH) are one and the same "Centaurus Great Wall" structure, in which our galaxy itself is embedded.

Acknowledgments. This research is supported by the Foundation for Research Development (South Africa).

References

Broadhurst, T.J., Ellis, R.S., Koo, D.C. & Szalay, A.S., 1990. Nature **343**, 726.

da Costa, L.N., 1992. 2nd DAEC Conference on The Distribution of Matter in the Universe, Edited by D. Gerbal & G. Mamon (Meudon, Observatoire de Paris Press), p163.

Fairall, A.P., 1993. Observational Cosmology, Edited by G. Chincarini, A. Iovino, T. Maccacaro & D. Maccagni, ASP Conference Series, **51**, 148.

Geller, M.J. & Huchra, J.P.,1989. Science **246**, 897.

Maurellis, A., Fairall, A.P., Matravers, D.R. & Ellis, G.F.R., 1990a. A&A, **229**, 75.

Maurellis, A., Ellis, G.F.R. & Fairall, A.P., 1990b. International Workshop on Superclusters and Clusters of Galaxies and Environmental effects, Sesto-Moso, Book of Abstracts.

Ramella, M., Geller, M.J. & Huchra, J.P., 1992. ApJ, **384**, 396.

Discussion

N. Lu: How uniform are your redshifts in different directions?

A. Fairall: As stressed earlier, the coverage using catalogued data is governed by the selection effects of individual observers. Nevertheless, one finds that many of the brighter galaxies (to around $m_B \sim 14.5$ or so) are repeatedly observed by different investigators, to the point where few, if any, are overlooked. Consequently there is reasonably uniform coverage (in direction) out to $cz = 10,000$ km s^{-1}. At much greater redshift, coverage is non-uniform since there is strong selection towards (the direction of) rich clusters.

T. Yamada: What is your definition of a 'wall' (not filament)? How many walls are seen in your catalogue? Did you do any statistical study of the walls in your catalogue (on their extensions, overdensities, etc.)?

A. Fairall: A 'great wall' is an obvious topographical feature, a large-scale concentration of galaxies within a volume about several thousand km s^{-1} long (in redshift space), a few thousand km s^{-1} wide and about a thousand km s^{-1} thick. In my catalogue, the Sculptor and Fornax Walls meet such a definition (as does the Great Wall and the Perseus-Pisces Wall in the northern skies). There are also lesser 'walls' more like sheets of galaxies - for which the Grus Wall is an example. Although we have examined certain properties and the content of the Grus Wall (Maurellis *et al* 1990b - see references above), statistical studies of overdensities etc. cannot be carried out with this data because it is not a controlled sample.

J. Huchra (comment to T. Yamada): A good statistical study of the distribution of galaxies in a wall, the Great Wall, can be found in a paper by Ramella *et al.* 1992 (listed above in References), where, if you look at the surface distribution of the galaxies in the wall, it is like swiss cheese, with a filling factor of about 1/3 and holes of a few hundred to 1000 km s^{-1} across.

D. Lynden-Bell: I just emphasize that this is uncontrolled data. If you concentrate in the direction of a cluster and look at everything, then you get a searchlight beam whose length has nothing to do with the velocity dispersion in the cluster itself.

A. Fairall: I agree!

H. van Woerden: Both in the 'slice slides' and in the movie, I was struck by a circular structure inside the Fornax Wall. Do you consider this a void, or is it only a larger 'foam' element? And what is the size of this structure?

A. Fairall: More than likely, this is what I have labelled the Sagittarius void, and its diameter is only 1000 km s^{-1} or less. This is very much smaller than the large voids (like the others labelled in Figure 2) and is therefore like the cellular structure that exists within great walls. Thus it supports my contention that the Fornax Wall and Centaurus supercluster together form a great wall.

THE LEIDEN/DWINGELOO SURVEY OF HI IN OUR GALAXY

W.B. BURTON AND DAP HARTMANN
Sterrewacht Leiden, P.O. Box 9513, 2300 RA Leiden, The Netherlands

ABSTRACT   The Leiden/Dwingeloo survey of HI emission from the Milky Way north of $\delta \geq -30°$, which has occupied the 25–m radio telescope of the Netherlands Foundation for Research in Astronomy for an observing run of about 5 years duration, is briefly described. The principal astronomical queries which motivated this undertaking are mentioned. A new long-term project, the Dwingeloo Obscured Galaxy Survey is introduced.

A survey of the HI sky north of $\delta \geq -30°$ has been carried out using the 25–m radio telescope of the Netherlands Foundation for Research in Astronomy, located in Dwingeloo. The spatial resolution of the new survey is determined by the HPBW of the antenna, which is 35.'2 or 0.°6 at 1420 MHz, and by the true–angle grid–spacing of 0.°5 in both l and b. The spectral resolution of the survey is determined by the spacing of 1.03 km/s between each of the 1024 channels of the DAS Dwingeloo Autocorrelator Spectrometer, developed by A. Bos. The useful kinematic range of the survey extends between velocities (measured with respect to the Local Standard of Rest) of –450 km/s and +400 km/s, and thus embraces the regime of high–velocity and intermediate–velocity clouds as well as the regime of galactic gas with conventional kinematic behaviour. The rms brightness temperature intensity sensitivity of the survey is about 0.07 K, and was achieved in 180 seconds of integration. The receiver used was similar to those used on the Westerbork Synthesis Radio Telescope. The system temperature was typically 35 K. The data have been corrected as indicated below for contaminating radiation entering the near– and far sidelobes of the antenna.

The observational parameters of the Leiden/Dwingeloo survey represent an improvement over those of the Heiles and Habing (1974) Hat Creek survey by about an order of magnitude in kinematic coverage and in sensitivity, and an improvement over those of the Stark *et al.* (1992) Bell Labs survey by about an order of magnitude in spatial and kinematic resolution. The material will be published by Cambridge University Press as an atlas of maps and as a FITS–format data cube on a CD–ROM (Hartmann and Burton, 1995). Figure 1 shows a representative (but at considerably poorer pictorial resolution than the CUP atlas) slice through the Leiden/Dwingeloo datacube.

The 0.07 K rms limit on the measured intensities is about an order of magnitude less than the peak intensity characteristically entering the far sidelobes of the Dwingeloo antenna; at lower $|b|$, it is typically two orders of magnitude less than the peak intensity entering the near sidelobes of the antenna. The rms sensitivity achieved clearly can only be exploited if the data are fully corrected for

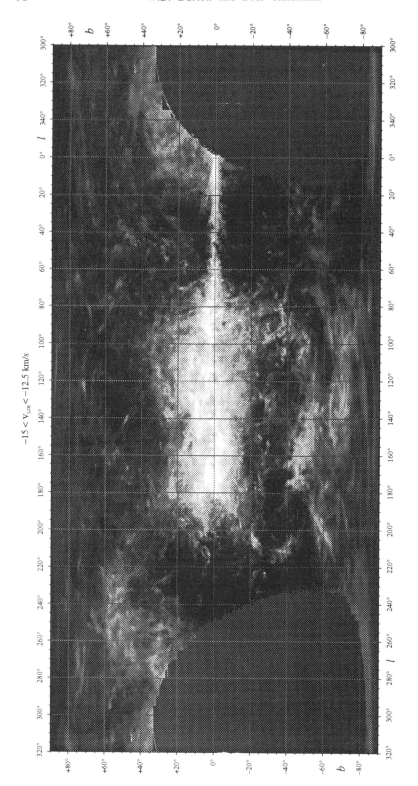

Figure 1. *Representative slice through the Leiden/Dwingeloo survey of HI in our Galaxy, showing the distribution on the plane of the sky of emission integrated over the velocity range −15.0 < v < −12.5 km/s.*

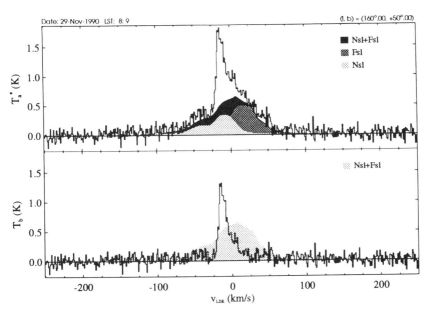

Figure 2. *Spectrum observed in the direction l=160°.0, b=50°.0, with the calculated contamination entering the near- and far sidelobes indicated separately in the upper panel. In the lower panel the corrected spectrum is displayed together with the total sidelobe contamination which was subtracted.*

contamination by stray radiation. The correction involved convolving the measured antenna pattern with the measured all-sky HI emission. Such a correction is by no means a trivial matter, as it requires extensive measurements of the antenna response to a strong source of continuum radiation as well as completion of the all-sky HI survey which then serves as input to the boot-strap correction. The algorithm used in application of the stray-radiation correction is the one developed by Kalberla (1978; see also Kalberla et al. 1982); the correction was applied in collaboration with P. Kalberla and U. Mebold at the University of Bonn.

Figure 2 illustrates the importance of the correction for stray radiation. In regions where the total HI emission is exceptionally weak (see e.g. Lockman et al. 1986) the total emission detected by the far sidelobes may be as strong as the emission from the direction of the main beam of the telescope. The antenna pattern of the far sidelobes (defined here as lying more than 16° from the axis of the main beam) detects emission from much of the sky. The amount of emission detected, and its velocity structure, depends principally on the perception of the Milky Way by the extended antenna pattern during the course of the observation. This perception varies fundamentally for different times and dates, and for different observed positions. When a single position is tracked, for example,

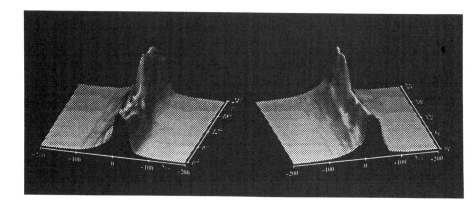

Figure 3. *Variation of the stray radiation entering the far sidelobes of the Dwingeloo telescope during a 24-hour period of LST when the telescope is pointed towards the direction $l=160°.0$, $b=50°.0$. The velocity scale runs from -200 km/s to $+200$ km/s; the intensity scale peaks between 0.41 K (LST=13:00) and 0.78 K (LST=17:00). The variations in the emission detected in the far sidelobes are due to changes in the amount and orientation of Milky Way emission as presented to the side-lobe pattern.*

different portions of the Milky Way are perceived as it rises and sets during the day and different portions will have varying relative motions with respect to the observed position.

Figure 3 shows the changing contamination from emission detected by the far sidelobes as the direction $l=160°.0$, $b=50°.0$ is tracked on a particular day over a 24-hour period of LST. The far-sidelobe contamination varies as the orientation of the Milky Way changes with respect to the antenna pattern of the (altitude–azimuth mounted) telescope; the extent of the Milky Way emitting above the local horizon also changes. The emission entering the far sidelobes is always a concern: because it is largely contributed by the HI disk of the galactic equator, its velocity structure typically extends over about 150 km/s. In directions of exceptionally low total integrated HI intensity the stray-radiation contamination may equal or exceed the true column depth; but even in regions of high intensity, correction for the far-sidelobe contribution is important because of its large velocity extent.

The near-sidelobe contribution in a given direction shows much less variation with time as the position of the main beam changes. The intensity of the near-sidelobe contamination can, however, be quite high: at low $|b|$, 10 K is a typical near-sidelobe contamination level, and, although this contamination contributes little additional kinematic structure to the profile in question, correction for it is particularly important to accurate determination of total column depths as well as to discussions of optical depths and physical temperatures.

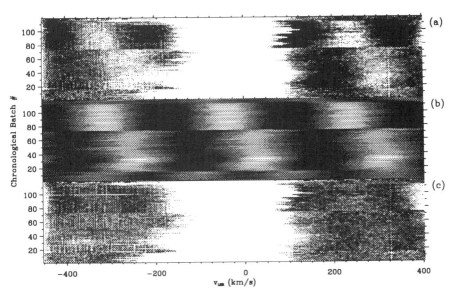

Figure 4. *Illustration of the correction for low-level ripples in the baseline structures due to standing-wave reflections. The upper panel shows the residual structure remaining in the averaged spectra corresponding to all observations in one 5° × 5° observing batch. The average was made after application of a polynomial baseline correction to all individual spectra in that batch. The middle panel shows the ripple as determined from a sine fit to each of the batch-averaged spectra; the amplitude of such sine fit is typically 15 mK, or about 0.2 rms. The lower panel shows the batch-averaged spectra after correction for the standing-wave ripple. The cause of the changes in phase which occurred on two well-defined dates has not been identified.*

The Leiden/Dwingeloo HI survey is suited to a range of astronomical investigations. Several of the studies which motivated the survey are briefly mentioned here (see also review by Burton, 1992).

The warp of the outer gaseous layer of the Milky Way has been studied in some depth, but the signature of the warp extends to latitudes and velocities beyond the reach of earlier survey material. The sensitivity and kinematic coverage of the new all–sky HI survey will allow an improved description of the shape of the outer Galaxy. It is particularly important to establish the dust content in the warped outer Galaxy; the detailed kinematic coverage of the high $|b|$ sky should allow separation of the local dust features, which are tightly correlated with the kinematic structures of local HI, from those contributed by interstellar material at large distances.

Several regions of exceptionally low total HI column depth have been identified, in addition to the one in Ursa Major studied by Lockman *et al.* (1986). These regions are important as low–extinction viewing ports to the extragalactic world. They are also important to discussions of turbulence and energetics of the gaseous disk and the lower halo of our Galaxy, and as regions where X–ray shadows cast by the HI gas may be studied.

The Leiden/Dwingeloo survey is the first to systematically encompass the kinematic range including the high– and intermediate–velocity clouds as well as the gas belonging to the conventional galactic disk. Relationships between the high–velocity and intermediate–velocity material will be investigated. The evident relationship of the IVC material with the conventional galactic disk gas will be given particular attention. Evidently dust cirrus features characterizing the IRAS survey generally have HI counterparts; it appears (see Deul and Burton, 1990, and Burton et $al.$, 1992) that many cirrus features have HI counterparts moving at velocities which are highly anomalous (compared to the kinematics which might be expected from a well–behaved galactic disk).

The motions and shapes of the structures characterizing the HI gas reveal important aspects of the macroscopic energetics of the interstellar medium. The extension of velocity information beyond the coverage and resolution available earlier shows that some of the structures, especially the large shell–like ones, can be traced over a larger kinematic extent than previously realized, suggesting an upward revision of the energetics involved. The topology and kinematics of the HI structures identified in the Leiden/Dwingeloo material will be investigated in terms of the areal filling factor. The kinematic resolution of the survey has revealed HI structures, isolated at high $|b|$, with exceptionally narrow velocity widths. There are only three molecular clouds known at high $|b|$ which also show anomalous ($|v| > 25$ km/s) kinematics, but each of these clouds has an associated narrow HI structure. A finding chart defined by narrow (dispersion < 1 km/s) HI features might lead to additional interesting molecular features.

An intended use of the Leiden/Dwingeloo HI survey which is appropriate to mention at a meeting on the Zone of Avoidance concerns the gas/dust/reddening queries investigated by Burstein and Heiles (1984). The material in the new HI survey, together with the modern IRAS and DIRBE dust data, are well suited to new analyses of some of the questions posed by Burstein and Heiles. A particularly puzzling aspect of the gas/dust relationship concerns the evident difference between the region of the Galaxy lying inward of the Sun's position (but including the local neighbourhood), where the correlation between HI and dust emissivities is quite tight, and the region of the Galaxy lying well outward of the Sun's position, where this correlation seems to break down.

With the completion of the Leiden/Dwingeloo survey of HI in our Galaxy, the telescope becomes available for another large–scale effort. The Dwingeloo Obscured Galaxies Survey (DOGS) aims to detect galaxies lying in the Zone of Avoidance within the redshift range to 4000 km/s. The project is a collaboration involving G. Hau, O. Lahav, A. Loan, and D. Lynden–Bell in Cambridge; R. Kraan–Korteweg in Groningen, W.B. Burton and Dap Hartmann in Leiden; H. Ferguson at the STScI; and P. Henning at the University of New Mexico. Several months of tests have verified the stability of the Dwingeloo system for long integrations, and have verified that the interference environment is suitably benign. The integration times of up to an hour per spectrum, and the bandwidth giving coverage of velocities over the range $0 < v < 4000$ km/s at 4 km/s resolution, will yield sensitivities of better than 0.01 K. A blind search will cover

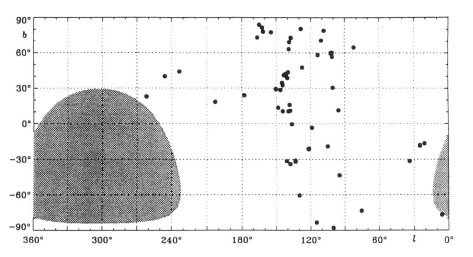

Figure 5. *Location on the sky of all external galaxies contaminating the Leiden/Dwingeloo survey (thus in the velocity range -450< v< +400 km/s). The exercise of detecting these systems was primarily important to gauging the quality of the baseline-fitting routine.*

regions of full obscuration, and a directed search will concentrate on candidate objects selected after scrutiny of optical and IRAS material.

Experience gained during the course of the Leiden/Dwingeloo survey of HI in the Milky Way will benefit the new obscured–galaxies project in a variety of ways. An example is given by the investigation of baseline ripples caused by standing waves bouncing back along the feed–support legs from the prime focus of the telescope. These standing waves result in low amplitude, broad ripples in the baselines. The signature of the ripples could have lead to confusion in identifying weak, high–velocity wings in the survey, and could have masked detection of nearby galaxies. Figure 4 illustrates that the baseline ripples could be accounted for by sine–curve fits.

Confidence in the baseline accuracy of the Leiden/Dwingeloo data is given by the detection of local external systems. Of course, the relatively short integration times of the Milky Way survey ruled out detection of galaxies other than bright ones well–known from previous, special–purpose work. Figure 5 shows the location of the galaxies entering the survey; Figure 6 shows representative spectra for a sample of these systems. We note that the largely automated baseline–determination routine identified these weak, broad systems naively. In the context of investigations of the Milky Way gas, this makes us quite confident that the baseline–fitting routine would not discriminate against similarly weak, broad spectral features contributed, for example, by high–velocity clouds.

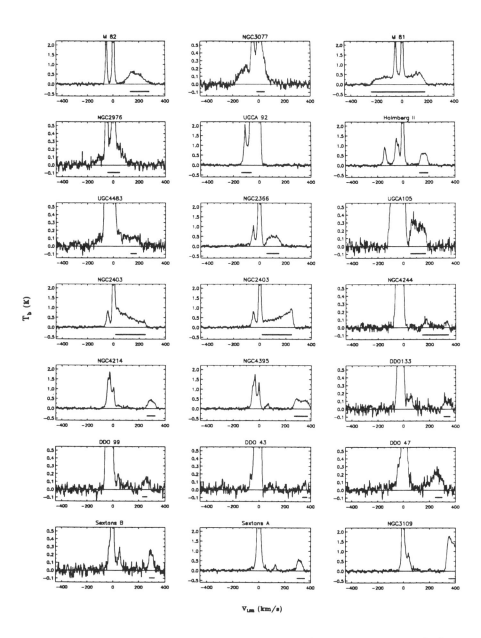

Figure 6. *Sample spectra from the Leiden/Dwingeloo survey which also contain emission from external systems.*

ACKNOWLEDGEMENTS

The Dwingeloo 25-m telescope is operated by the Netherlands Foundation for Research in Astronomy with support from the Netherlands Organization for Scientific Research (NWO). A graduate research fellowship from NWO supported Hartmann's participation in this project; Burton's attendance at this meeting was supported by a grant from the Leiden Kerkhoven–Bosscha Foundation. The stray–radiation correction was applied using the resources of the Radio Astronomical Institute of the University of Bonn, in collaboration with P. Kalberla and U. Mebold.

REFERENCES

Burstein, D., & Heiles, C. 1984, ApJS, **54**, 33

Burton, W.B. 1992, in *Saas–Fee Advance Course* 21, "The Galactic Interstellar Medium", D. Pfenniger & P. Bartholdi (eds.), Springer–Verlag, pp. 1–155

Burton, W.B., Bania, T.M., Hartmann, Dap, & Yuan, T., 1992, in *Proceedings CTS Workshop* No. 1, "Evolution of Interstellar Matter and Dynamics of Galaxies", J. Palouš, W.B. Burton, and P.O. Lindblad (eds.), Cambridge University Press

Deul, E.R., & Burton, W.B. 1990, A&A, **230**, 153

Hartmann, Dap 1994, *"The Leiden/Dwingeloo Atlas of Galactic Neutral Hydrogen"*, Ph.D. Dissertation, University of Leiden (in preparation)

Hartmann, Dap, & Burton, W.B. 1995, *"The Leiden/Dwingeloo Survey of Neutral Atomic Hydrogen in the Galaxy"*, Cambridge University Press (under contract)

Heiles, C., & Habing, H.J. 1974, A&AS, **14**, 1

Kalberla, P.M.W. 1978, Ph.D. Dissertation, University of Bonn

Kalberla, P.M.W., Mebold, U., & Reif, K. 1982, A&A, **106**, 190

Lockman, F.J., Jahoda, K., & McCammon, D. 1986, ApJ, **302**, 432

Stark, A.A., Gammie, C.F., Wilson, R.W., Bally, J., Linke, R.A., Heiles, C., & Hurwitz, M. 1992, ApJS, **79**, 77

Discussion

E. Valentijn: I am impressed by the quality of your data, particularly by the strong correlations between the HI and 100 μm maps. You also point out the much stronger decrease of 100 μm emissivity than HI emissivity at large radii. The contrast between these two observations is puzzling and perhaps one of the most fundamental problems. We do observe a similar HI/IRAS 50 μm variation in external galaxies. It seems therefore that a comparison of your results with scale length ratio's $\left(\alpha_K, \alpha_B, \alpha_{50\mu m}, \alpha_{100\mu m} \right)$ of external galaxies might help to resolve this important issue.

Unveiling Large-Scale Structures Behind the Milky Way
ASP Conference Series, Vol. 67, 1994.
C. Balkowski and R. C. Kraan-Korteweg (eds.)

THE 2 MICRON ALL-SKY SURVEY

J. Huchra

Center for Astrophysics, 60 Garden Street, Cambridge, MA 02138, USA

W. Pughe, S. Kleinmann, M. Skrutskie, M. Weinberg

Department of Physics and Astronomy, University of Massachusetts, Amherst 01003-4525, USA

C. Beichman, T. Chester

IPAC, Caltech, 770 South Wilson St. Pasadena, CA 91125, USA

Abstract. The 2 μm All Sky Survey, or 2MASS, will cover the sky at three near-infrared wavelengths (J, H, and K_s) with a pixellation of $\sim$2.0" One of the primary science goals of the survey is to produce a catalogue of galaxies to a limit between K=13 and K=14 (equivalent to B=16-17) to within 10° of the galactic plane. The 2MASS Galaxy Catalogue is expected to contain over a million galaxies and will be an invaluable aid to mapping the local density field out to z=0.1.

1. Introduction

One of the major current problems in cosmology is the lack of understanding of the cause/source of our motion with respect to the cosmic microwave background radiation (e.g. Smoot *et al.* 1991). This lack of understanding is evidenced by the large number of often contradictory results on the local flow field (Lynden-Bell *et al.* 1988; Scaramella *et al.* 1991; Lauer and Postman 1994). Attempts at resolving this issue *via* the combination of maps of the velocity field and of the galaxy density field so far have been hampered by either the sparseness of the samples, their lack of depth, and/or their lack of sky coverage. The IRAS QDOT (Rowan Robinson *et al.* 1990) and 1.2-Jy (Strauss *et al.* 1992) samples, while having good sky coverage, do not densely sample the field and contain only spiral galaxies, which are known to be biased away from high density regions. The optical surveys (Lahav and Scharf 1993), while less biased by morphological type, and more dense (e.g. Geller and Huchra 1989; Marzke *et al.* 1994), are significantly affected by galactic extinction. Reconstruction of the density field from the velocity field and *vice versa* has only been done on small scales and with relatively crude ($\sim$1000 km/sec smoothing scales) spatial resolution (Bertschinger and Dekel 1989; Dekel *et al.* 1993). Perhaps one of the largest difficulties with these attempts is that the direction of the microwave background motion, after correction to the centroid of the Local Group, is only 30° from the galactic plane.

In order to address this problem more directly (as well as to provide good samples to attack a broad spectrum of galactic and extragalactic astrophysical problems), we are about to conduct a digital all-sky survey in the near-infrared in three bands. The advantages of such a survey for mapping the local galaxy density field are obvious. Firstly, the survey uniformity down into the galactic plane is considerably improved over optical surveys because $A_K = 0.07A_B$. Secondly, a near-IR survey is much more "morphology blind" than either optical or IRAS surveys. Galaxy energy distributions for almost all types of galaxies peak near 1.6 μm (Aaronson 1978). In addition, near-IR wavelengths sample the stellar mass distribution much better than blue or 60 μm wavelengths, which can be (are!) dominated by young stellar components that represent only a small fraction of a galaxy's mass. Remember that the primary relation that we are attempting to apply while mapping large scale flows and density fields is the fact that both light and gravity fall off as $1/r^2$ — if the "light" we are measuring is only weakly related to a galaxy's mass (e.g. 60 μm radiation which comes primarily from hot dust), then perhaps it is not fair to use samples selected in such a manner to provide definitive cosmological answers.

Lastly, providing uniform photometry for a large sample of objects is absolutely necessary because small errors in photometric zero points can translate into apparent density variations that are significant on the clustering scales we need to examine (e.g. de Lapparent, Kurtz and Geller 1986). Photographic surveys have well known zero point and linearity problems. In addition, a digital survey such as 2MASS, can provide accurate magnitudes for individual galaxies for use in applications like the infrared Tully-Fisher relation. We intend to use these for the measurement of the Hubble constant *and* to better determine the very galaxy velocity field we are trying to understand through the density field.

2. SURVEY IMPLEMENTATION

The critical parameters for 2MASS are listed in Table I. The survey will operate simultaneously in two hemispheres (to achieve full sky coverage) and in three near-infrared wavebands. The observations will be carried out with a pair of 1.3m telescopes which will be optimized for the needs of the survey, and are fully dedicated to the survey execution. Each telescope will be equipped with a camera containing three infrared array detectors.

The arrays will record images with a pixel scale of 2.0"/pixel. The integration time is set so that the observations are background-limited in all three wavebands; OH airglow dominates the background in the J and H bands, while thermal emission from the telescope also contributes to the K_s-band background. The survey specifications call for the detection of point sources with $K_s \leq 14.0$ mag. at $\geq 10\sigma$.

A novel mapping strategy, dubbed "freeze-frame scanning" by Frank Low, is used to scan the sky. In this scheme, the telescope tracks in Right Ascension, while executing a smooth scan in Declination. The secondary mirror is moved in a sawtooth pattern, such that during the slow sweep of the secondary, the image of the sky is frozen onto the focal plane. At the end of the slow sweep (which typically takes about 1.5s) the secondary is moved quickly back to its starting point, but the telescope is pointing to a new location. The scan rate is chosen such that the new location is displaced only 1/5th of the field from the previously imaged field, so that 5 separate images are obtained for each field on the sky. The array is slightly tilted with respect to the scan direction, to improve the intra-pixel sampling. This scheme minimizes the effects of bad pixels, and improves both the photometric and positional accuracy.

To test the survey mapping strategy, measure the backgrounds, and determine some critical astrophysical parameters needed to optimize the design of the survey, a prototype camera was built in 1992 at Infrared Laboratories. This camera contains only a single, 256x256 HgCdTe array, and therefore maps in only 1 band at a time. The prototype camera is capable of surveying the sky at a rate of 13 square degrees per hour. This implies a required survey duration of approximately 2000 photometric hours to map a hemisphere or between 500 and 1000 nights of time, 2-3 years, if weather is factored in.

Table 1.   2MASS Parameters

	Survey Parameters	Prototype Camera
Wavebands	J, H, and K_s	J, H, or K_s
Sky Coverage	$4\,\pi$ sr	800 sq. deg. (to date)
Pixel Size	$2.0'' \times 2.0''$	$2.3'' \times 2.3''$
Sensitivity	$K_s \leq 14$ mag., 10σ	$K_s \leq 14$ mag., 14σ
Photometric Accuracy	$\leq 5\%$ for $K_s \leq 12.5$ mag.	$\leq 5\%$ for $K_s \leq 12.5$ mag.
Positional Accuracy	$\leq 1.0''$	$\sim 0.5''$

3. PROTOTYPE CAMERA RESULTS

We have used the prototype camera to survey several dozen areas of the sky, including regions around the Orion Molecular Cloud, Coma, and a section of the galactic plane at b=0° and l=53° (19^h27^m $+17°42^m$ where we have also digitized the Palomar B and R plates for comparison. This comparison is shown in Figure 1. It is easily seen that the 2MASS scans, even though short in total integration time, probe considerably deeper in and/or through the plane than optical photographic surveys.

One issue that has been raised is the ability to detect galaxies at low galactic latitudes with our moderately coarse pixellation. The initial survey of the plane is likely to be done with 2″ pixels. While it is difficult to know exactly what our limits will be as a function of galactic latitude, we can get some qualitative feeling by looking at one or two well known objects. Figure 2 shows our survey mode image of Maffei 2. This can be compared to your favorite optical image of the galaxy.

A test of the ability of 2MASS to detect galaxies at high galactic latitude was performed by observing the core of the Coma cluster. Figure 3 shows the survey mode K_s-band image. Comparison of this image with the POSS-E (red)

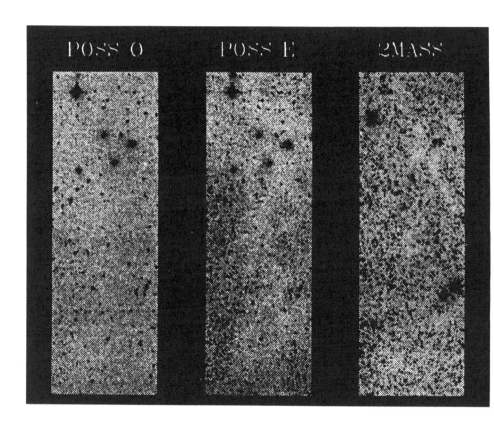

Figure 1. North-South scans of a region of the Galactic plane at b=0°, l=53°, between 17°.32′< δ < 17°.55′ and 19ʰ.26.7ᵐ< α < 19ʰ.27.5ᵐ.

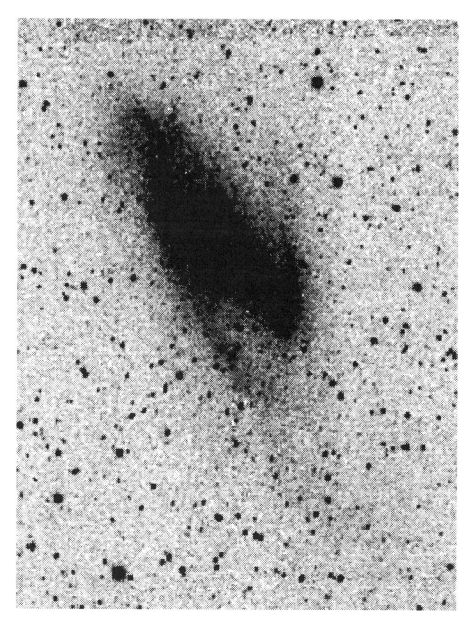

Figure 2.   Our K_s-band survey mode image of Maffei 2.

print shows that all of the brighter optically detected galaxies are easily visible on the 2μm image. This is extremely impressive considering the ratio of exposure times in the images—about 45 minutes for the POSS plate and 7.5 seconds for the 2MASS scan.

Tests of stellar photometry have also been done and provide a benchmark for the detection limits for point sources. In the configuration described above, 2MASS will reach a S/N of 10 for stars of J $\leq$ 15.3, H $\leq$ 15.1, and K$_s$ $\leq$ 14.3.

4. THE ABELL 262 FIELD

Perhaps the most critical experiment that we have performed with respect to the detection and selection of galaxies is the analysis of an 11 square degree field near Abell 262. The preliminary results presented here were derived by applying the FOCAS algorithm to extract nonstellar objects from the 2MASS images. FOCAS was developed by Tyson and Jarvis (1979; see also Jarvis and Tyson 1981) to count faint galaxies on photographic plates, and has since become one of the standard image classification algorithms and is implemented in IRAF. We are currently trying additional techniques. Here, we present some preliminary results from the A262 field.

Over 200 objects were found having K$_s$ $\leq$13.5. Figure 4 compares the distribution of our K$_s$-selected objects to optically selected galaxies from the Zwicky catalogue (1961-8). Essentially all the objects in the Zwicky catalogue are detected in K$_s$ to a limit of 13.5. At that limit there are approximately 4 times as many K$_s$- selected objects as B selected objects. We detect 30, 48, 86 and 220 galaxies to K$_s$ limits of 11.5, 12, 12.5 and 13.5, respectively.

For those galaxies detected in both samples, we can compare the B and K$_s$ magnitudes. Figure 5 is a plot of B$_{Zwicky}$ versus K$_s$. The Zwicky magnitude scale has been checked by several authors (e.g. Huchra 1976). The approximate linearity of the plot as well as the near unit slope is an indication that the magnitudes are moderately well measured. The scatter is representative of the range in B-K$_s$ colors of galaxies. The average B-K$_s$ is ~3.5, consistent with the known optical-IR colors of galaxies (Aaronson 1978).

We also have measured redshifts for 52 of the 86 galaxies brighter than K$_s$=12.5 (and will measure redshifts for the fainter galaxies this fall). The mean absolute K$_s$ magnitude, M$_{K_s}$, is -24.3 (for H$_0$ = 50 km/s/Mpc) or -22.9 (for H$_0$ = 100). These numbers, too, are consistent with a mean B-K$_s$ = 3.5 and a characteristic luminosity, M$_B^*$ = -19.4 (deLapparent, Geller and Huchra 1989).

We can also use the overlapping scans in this region to determine the *internal* error in magnitude determination. Figure 6 shows the cumulative magnitude difference distribution between multiple scans. The two lines are for different detection thresholds. For a 3σ threshold, the distribution is indicative of a scatter of only 0.11 magnitude. At a lower detection threshold, 2.5σ, the derived scatter rises, as expected, to 0.18 magnitude.

5. SUMMARY

Although preliminary, these results confirm our ability to detect galaxies to limits significantly deeper than existing optical surveys. More work is needed

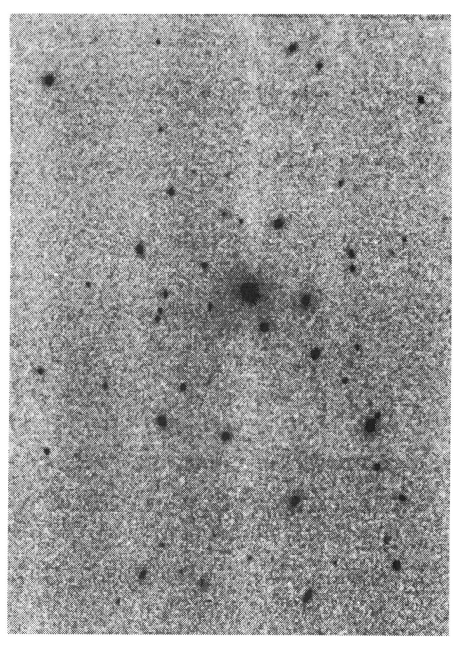

Figure 3.   The K_s-band survey mode image of a central region of the Coma cluster of galaxies.

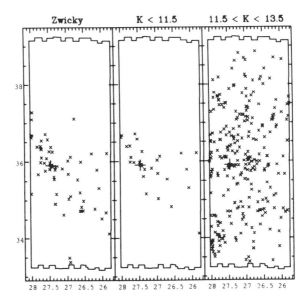

Figure 4.   *Left:* All galaxies in the A262 scans with $B \leq 15.7$ from
the Zwicky catalogue. *Center:* All galaxies with $K_s \leq 11.5$ in the
same region. Note the similarity in the surface distribution of these
bright galaxies. There are approximately 50% as many galaxies to
$K_s = 11.5$ as to $B = 15.7$. *Right:* Galaxies with $11.5 < K_s < 13.5$. There
are about 4 times as many K_s-selected galaxies to this limit than in the
Zwicky catalogue. The fainter galaxies are also much more uniformly
distributed on the sky indicating that they are likely background to
the A262 cluster.

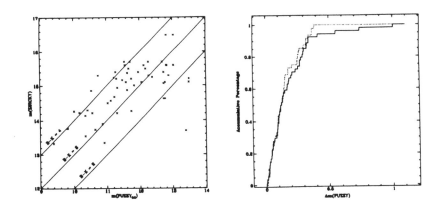

Figure 5. Zwicky magnitude versus K_s-band magnitude for jointly selected galaxies.

Figure 6. The cumulative distribution of magnitude differences in overlapping scans. The solid line is the result from 2.5σ detections, the dotted line is for 3.0σ detections.

to determine the true completeness limits of the 2MASS survey, but it appears as if we are easily complete to at least $K_s = 12.2$ from the comparison with the optical catalogs. If our measured galaxy surface density to $K_s = 13.5$ in the A262 region is representative—220 galaxies in 11 square degrees—2MASS will detect and catalog over 500,000 galaxies. If we can combine the JHK_s scans to produce a deeper catalog, the number will be much higher. In either case, we expect 2MASS to be an outstanding tool for probing the galaxy distribution through the Zone of Avoidance.

References

Aaronson, M. 1978,ApJ, 221, 103.

Bertschinger, E. and Dekel, A. 1989,ApJ,336,L5

Dekel, A. Bertschinger, E. Yahil, A., Strauss, M., Davis, M. and Huchra, J. 1993,ApJ,412,1

de Lapparent, V., Kurtz, M. and Geller, M. 1986,ApJ,304,585

de Lapparent, V., Geller, M. and Huchra, J. 1989,ApJ,343,1

Geller, M. and Huchra, J. 1989,Science,246,897

Huchra, J. 1976, AJ,81,952

Jarvis, J. and Tyson, J. A. 1981, AJ, 86, 476.

Lahav, O. and Scharf, C. 1993, MNRAS, 262, 711.

Lauer, T. and Postman, M. 1994, ApJ, in press.

Lynden-Bell, D. et al. 1988,ApJ,326,19

Marzke, R. et al. this conference.

Rowan-Robinson *et al.* 1990. MNRAS, 247,1.

Scaramella, R., Vettolani, G., and Zamorani, G. 1991, ApJ, 376, L1.

Strauss, M., Huchra, J. P., David, M., Yahil, A., Fisher, K. B., and Tonry, J. 1992. ApJS, 83, 29.

Tyson, J. A. and Jarvis, J. 1979, ApJ, 120, L153.

Zwicky, F., Herzog, E., Wild, P., Karpowicz, M. and Kowal, C. 1961-8, *Catalogue of Galaxies and of Clusters of Galaxies*, (Pasadena: Caltech).

Discussion

G. Mamon: How far can you resolve spiral structure in your prototype camera images?

J. Huchra: We can roughly see features to a limiting K band surface brightness of 20.0 - 20.5. This corresponds to a surface brightness in B at $\sim$ 24, which is a little smaller than the Holmberg limit.

P. Felenbok: How many IR objects do you have per square degree?

J. Huchra: As you saw, we have $\sim$ 220 galaxies in the 11 square degree region around Abell 262 to K=13.5, or about 20 galaxies per square degree. This is what lead to my estimate of $\sim$ 500,000 galaxies over the sky to the survey limit.

N. Lu: To what magnitude limit, is the 2 MASS photometry good enough so that a new Tully-Fisher relation analysis using 2 MASS K-band data is worth trying?

J. Huchra: We expect to be able to do better than 10% absolute photometry to K$\sim$12.5 for most galaxies, but remember that because of the steep slope of the IRTF (and, indeed, most luminosity-velocity dispersion measurements) most of the distance errors will be due to linewidth measurement errors or the natural dispersion in the relation - 10% photometry $\rightarrow$ 5% distances if the error is in the photometry alone.

O. Lahav: What will be your strategy for a redshift follow-up of the 2 μm two-dimensional catalogue?

J. Huchra: Initially we will probably restrict ourselves to a sample limited to $\leq$ 13.5 at K of $\sim$ 200,000 galaxies. In the north we will use such telescopes as the Mt. Hopkins 60-inch. With improved detectors now being used we should be able to get the 50,000 or so remaining galaxies in $\leq$ 10 years. In the south we will require the good auspices of Tony Fairall!

C. Balkowski: Do all the galaxies you detected in the K band in A262 cluster (and for which you have the redshift) belong to the cluster?

J. Huchra: You can answer that by looking at the sky distribution as a function of magnitude. Most of the galaxies brighter than K=11.5 - 12.0 are at $\sim$ 4000 km/s and are members of A262 on the Pisces-Perseus chain, but most of the fainter objects near 13.5 are probably background. We only have redshifts for 2/3 of the bright ones.

W. Saunders: Over what fraction of the sky will you be able to compile a reliable galaxy catalogue?

J. Huchra: With the survey parameters I described (2" pixels) we expect to be reasonably complete to K=13.5 to about $| b | \sim 5°$. If time and funding (!) allow, we will resurvey the plane $| b | < 5°$ with better pixellation (1" pixels), with which we should be able to do a reasonable job given confusion to within a few degrees of the plane.

Unveiling Large-Scale Structures Behind the Milky Way
ASP Conference Series, Vol. 67, 1994.
C. Balkowski and R. C. Kraan-Korteweg (eds.)

How well should the DENIS Survey probe through the Galactic Plane?

Gary A. MAMON

*Institut d'Astrophysique, 98 bis Blvd Arago, F-75014 Paris, FRANCE
& DAEC, Observatoire de Paris-Meudon, F-92195 Meudon, FRANCE*

Abstract. The DENIS 2 micron survey is described. It's ability to probe galaxies at very low galactic latitudes is computed using simulated images. This analysis indicates that DENIS will be able to separate galaxies from stars, right through the Galactic Plane, for galactic longitudes $45° \leq \ell \leq 315°$, with a maximum loss in the magnitude limit for reliable extraction of $\simeq 1.2$ magnitude relative to that at the Galactic Pole (where the 92% reliability limit is $K' = 13.2$ for star/elliptical galaxy separation, and fainter for other morphological types). Extinction corrected magnitude limits are substantially worse, but for $|b| \geq 2°$, they are within one magnitude from that of the Galactic Poles. Thus, confusion by stars will not prevent the establishment of a reliable galaxy catalog right through the Galactic Plane. However, extinction will produce galaxies of lower surface brightness, which will be more difficult to detect.

1. Introduction

The recent advent of large 2D detectors, sensitive to light in near-infrared wavelength bands, holds great promise for extragalactic astronomy. Indeed, our view of the Universe is hampered by the dust layer in the Galactic Plane: in optical wavebands this dust is optically thick, and the optical detection of galaxies is severely reduced at galactic latitudes $|b| < 5°$ (*e.g.*, Weinberger, Woudt, Kraan-Korteweg, all in these proceedings). In the far infrared, probed by the IRAS satellite, this cool dust emits thermally and causes confusion with external galaxies, which becomes serious for $|b| < 12°$ within 90° from the Galactic Center (Meurs & Harmon 1988; see also Meurs and Saunders, both in these proceedings).

Similarly, our view of external galaxies is hampered by extinction from their interstellar dust in the optical, and thermal emission by this dust in the far infrared.

Moreover, both the optical (especially blue) and far-infrared luminosities of galaxies are enhanced by recent star formation, and it is thus believed that the near infrared light correlates best with the stellar content of galaxies, independent of morphological type (see Jablonka & Arimoto 1992).

With these advantages in mind, various members in the extragalactic community in Europe joined DENIS (*DEep Near-Infrared Southern Sky Survey*), while many of our American colleagues joined the very similar 2MASS (*2 Mi-*

cron All Sky Survey) project (see Huchra, in these proceedings). These two surveys should provide detailed mapping of the local Universe, without being biased in favor of star forming galaxies. The applications of near-IR surveys to the study of groups and clusters of galaxies have been reviewed by Mamon (1994). The aim of the present work is to check how well will the DENIS survey probe large-scale structure of the Universe behind the Galactic Plane.

2.   Overview of DENIS

The DENIS survey will map the entire southern sky ($-90° \leq \delta \leq +2°$) in the I (0.8μm), J (1.2μm), and K' (2.1μm) wavebands. The observations will make use of the existing ESO 1m telescope, on a 2/3 time basis until August 1994, and in principle full-time after that. The survey is expected to begin around the end of 1994 and last 3 years. Rapid mapping imposes a large field of view (12′), which amounts to 3″ pixels for the J and K' bands, using the current state-of-the-art 256×256 HgCdTe NICMOS-3 arrays designed by Rockwell, and 1″ pixels in the I band using a Tektronix 1024×1024 CCD.

The sky will be scanned in declination along strips, 30° long and 12′ wide, in a stop-and-stare mode. Elementary images of 12′ × 12′ are integrated for 9×1 s in J and K', making use of a micro-scanning device to dither the images by 1/3 pixel, yielding a pseudo-resolution in J and K' of 1″.

Preliminary calculations indicate that the DENIS survey will detect roughly 10^8 stars ($K' < 14.5$) and 250 000 galaxies ($K' < 13.7$), the latter with reliable star/galaxy separation and K'-band photometry accurate to 0.2 magnitudes (Harmon & Mamon 1993).

The raw data will be reduced at the Paris Data Analysis Center (PDAC), then shipped to the Leiden Data Analysis Center (LDAC) for the extraction of point sources and the bright galaxies, while the faint galaxies will be extracted back at the PDAC.

3.   New image simulations

The early image simulations presented by Harmon & Mamon (1993) took into account neither the range of morphological types of galaxies, nor the confusion with stars and extinction at low galactic latitudes. These issues are addressed here.

Artificial images are simulated by placing galaxies, of a given morphological type (*Ellipticals, Edge-On Disks*, and *Face-On Disks*), on a square grid, where in each row, galaxies have the same total magnitude. Stars are overlaid at random positions, according to a home-built star-count program, similar in spirit to that of Bahcall & Soneira (1980). An image is built using the IRAF ARTDATA package, taking into account the parameters of the atmosphere, telescope, and detector. A Moffatt PSF ($\beta = 2.5$) is assumed.

For the J and K' bands, the 9 subimages for each exposure are interlaced into a single image, with 768×768 1″ pseudo-pixels, with the information on 1 pseudo-pixel taken from that of *the* concentric 3″ pixel, which comes from a single subimage. In fact, images twice the size of the DENIS images (1536×1536 pixels of $1''$) are used, for increased statistics (grids of 20×20 galaxies). The image is saved in 16-bit format as in the survey, with the same sampling ($\sigma = 2\,\mathrm{ADU}$). The input stellar luminosity functions in the I and J bands are taken from Mamon & Soneira (1982). The K-band stellar luminosity function is adapted from Wainscoat *et al.* (1992), which appears to provide significantly better fits to existing star count data than the K luminosity function of Mamon & Soneira. Its parameters, in the notation of Bahcall & Soneira and Mamon & Soneira, are $\alpha = 0.40$, $\beta = -0.01$, $1/\delta = 1.0$, $M* = 3.30$, and $n* = 0.014\,\mathrm{mag}^{-1}\,\mathrm{pc}^{-3}$. The random stars are placed down to a magnitude limit 1 magnitude fainter than the theoretical magnitude limit for $5\,\sigma$ detection of point sources, so that the faintest stars contribute to increased noise in the diffuse background.

Rather than assume, as Bahcall & Soneira (1980), a plane parallel absorbing sheet, we adopt an absorbing layer that has an exponential distribution both in height (scale-height $= 100\,\mathrm{pc}$) *and* in length (scale-length equal to the stellar scale-length of $3.5\,\mathrm{kpc}$). We fix $A_V = 0.15$ to infinity at the Galactic Pole and $A_V = 30$ from the Sun to the Galactic Center (assumed to be $8\,\mathrm{kpc}$ distant). Because the absorbing layer is very heterogeneous, we will compare, in the future, with the Burstein & Heiles (1982) maps of galactic extinction for $|b| > 10°$.

The bulge (or spheroid) of our Galaxy is assumed to follow Young's (1976) deprojection of the $r^{1/4}$ law, with axis ratio 0.9, and effective radius of $8\,\mathrm{kpc}$. The local disk to bulge number density ratio is chosen as 800.

Elliptical galaxies are modeled as $r^{1/4}$ laws, while disk galaxies are exponential disks. The extrapolated central unattenuated blue surface magnitude is $\mu_B(0) = 14.0$ for ellipticals and 21.3 for disks. After determining the half-light radius, the galaxies are attenuated by the extinction to infinity returned by the star count program.

Table 1 below lists some of the DENIS parameters used in these simulations

Table 1.   Input Parameters

	I	J	K'
$\lambda\ (\mu)$	0.8	1.2	2.1
A_λ/A_V	0.48	0.28	0.11
μ_sky (mag arcsec^{-2})	19.0	16.0	12.5
Seeing ($''$ FWHM)	1.2	1.1	1.0
Quantum efficiency	0.60	0.60	0.65
Optical transmission	0.35	0.30	0.30
Read-out noise (e^-)	35	50	50
Integration time (s)	7.5	9×1	9×1
Pixel size ($''$)	1	3	3
Disk color ($B - C$)	1.4	1.9	2.8
Bulge color ($B - C$)	2.5	3.2	4.2

$B - C$ is the color $B - I$, $B - J$, $B - K'$, for bands I, J, and K', respectively.

4. Galaxy Extraction

There are various steps in extracting galaxies from images:

4.1. Detection

One wants to have a *complete* sample, in terms of a high probability that given an object, it is detected. One wants to have a *reliable* sample, so that a detected object is not caused by noise. Recent work of ours (Mamon & Contensou 1993) has convinced us that the interlaced images must be *smoothed* to detect galaxies, which are low surface brightness objects, in comparison with stars. The smoothing filter could be a boxcar, gaussian, or something else. It turns out that there is little difference in performance between boxcar and gaussian filters *in high galactic latitude fields*, but we have not yet tested the differences in crowded fields, for which galaxies will often be blended with stars. This is an important issue because in low star density regions, the smoothing scales need to be roughly 3 to 4 times greater to optimally extract the face-on disk galaxies, than to optimally extract the edge-on galaxies, the ellipticals, and the stars themselves. We hope to gain by first removing the stars detected by the LDAC. Note that a 3×3 boxcar smoothing amounts to transforming the interlaced image into an image where each $1''$ pixel corresponds to the average of the 9 $3''$ pixels containing that $1''$ pixel (one per subimage).

4.2. Star-Galaxy Separation

We wish to have a reliable star/galaxy classifier, especially in the sense that objects called galaxies will not in fact be stars. This is complicated by the fact that the ratio of star to galaxy number densities ($K' < 14$) varies from 10 to $> 10^4$ from the Galactic Poles to the Galactic Plane.

A simple star-galaxy separation algorithm is applied to each object in the *unsmoothed* image. The algorithm computes the mean surface brightness above a threshold for pixels within a given circular aperture around the true center of the object. This statistic should discriminate well between galaxies and blended stars. The aperture is set to be small, to avoid blending in with neighboring stars, although not too small to avoid noisy statistics. The threshold is a few background standard deviations above the background.

4.3. Photometry

For most cosmological applications, we wish to have a galaxy sample with accurate photometry ($\Delta m \leq 0.2$). We have performed tests of isophotal photometry, which show that the mean offset between isophotal magnitude and total magnitude depends on the isophote, the type of the object, and sometimes its magnitude. Presumably, this mean offset can be modeled. The photometric accuracy will be the combination of the dispersion of the offset and the uncertainty on its mean. These issues are still under investigation. In crowded fields, photometry will be biased by neighboring stars. The hope again is to gain by first removing the stars detected by the LDAC.

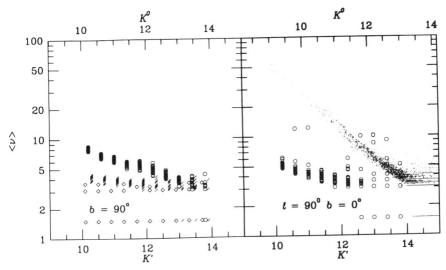

Figure 1.   Star galaxy separation in the K' band: mean surface brightness above threshold versus magnitude.   Ellipticals, edge-on disks, face-on disks, and stars are shown as *circles*, *dashes*, *diamonds*, and *points*, respectively. The upper scale is the equivalent K' magnitude for elliptical galaxies, after correction for extinction. The symbols at $\langle \nu \rangle = 1.5$ are objects for which no pixels were above the threshold.

5.   Results on Star/Galaxy Separation

Figures 1, 2, and 3 plot the star/galaxy separation statistic (in units of the true background standard deviation) as a function of true object magnitude for the Galactic Poles (left-hand plots) and for $\ell = 90°$, $b = 0°$ (right-hand plots). Recall that the stars are field stars with the predicted number in each magnitude bin, while the galaxies are test galaxies (at an expected 20 galaxies per square degree at $K' < 14$, one expects 4 galaxies by quadruple-area DENIS like field, hence galaxy-galaxy confusion will be negligible). The galaxies that are above the galaxy locus are blended with stars, and the ordinate of such a galaxy indicates the surface brightness, hence magnitude of the star with which it is blended. The objects are assumed to have already been detected, and that their astrometry and photometry are relatively accurate. In the next step (in preparation), we plan to relax these assumptions by doing a global pass at detection, star/galaxy separation, astrometry and photometry.

The figures indicate that stars are more concentrated than galaxies, especially at bright magnitudes. At bright magnitudes, the mean surface brightness of stars scales as $\mathrm{Cst} + 2.1\,m$, where m is the magnitude (one expects a slope of 2.5 at even brighter magnitudes). The limit for 92% reliable star/galaxy separation is $K' \simeq 13.2$ (see also Harmon & Mamon 1993) for ellipticals. The limits in J are slightly better, thanks to the lower background, while in I, the limits are substantially better thanks to the increased angular resolution.

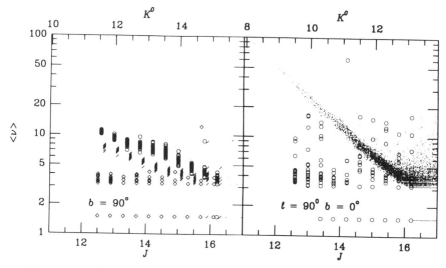

Figure 2.   Same as Fig. 1, but in the J band.

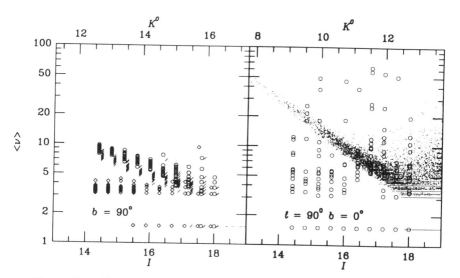

Figure 3.   Same as Fig. 1, but in the I band.

The elliptical galaxies are more concentrated than the edge-on disk galaxies, which themselves are more concentrated than face-on disk galaxies. The 92% reliability limits are $K' \simeq 13.6$ for edge-on disks, and $K' \simeq 13.7$ for face-on disks. However, less than half of these latter-type galaxies will be detected at these magnitudes. Although they are hardest to detect, face-on disk galaxies, once detected, are the easiest to tell apart from stars.

The right-hand plots in Figures 1, 2, and 3 show the effects of confusion and extinction in the Galactic Plane. The stellar locus is unaffected by extinction (it just widens with increased stellar density), *i.e.*, the mean surface brightness of unresolved point-sources is a function of their magnitude and the point-spread-function. On the other hand, with higher extinction, the mean surface brightness of galaxies tends to decrease as $\mathrm{dexp}(-0.4\,A_\lambda)$. So, the galaxy locus is, in fact, better separated from the stellar locus in high extinction fields (but, for the same reason, galaxies will be more difficult to detect). However, in the K' and I bands, star/galaxy confusion brightens substantially the 92% reliability limiting magnitudes. This is much less so the case in the J band, for which confusion always sets in at about the same magnitude as where the galaxy locus runs into the stellar locus for $|b| = 90°$ (see Table 2 below).

Table 2 provides a summary of the results of star/galaxy separation with the simple algorithm described above. One sees that, over most of the sky ($|b| \gtrsim 6°$), the I band, thanks to its 3-fold increased angular resolution, provides superior star/galaxy separation, such that it will be able to separate galaxies that would not be detectable in K' from stars. At faint magnitudes, the I band is read-out noise limited. If the read-out noise is chosen to be $50\,e^-$ instead of 35, then the magnitude limits, in Table 2, should be decreased by about $2.5\log_{10}(50/35) \simeq 0.4$ magnitudes, and the I band provides the best star/galaxy separation at high galactic latitudes for a read-out noise $\lesssim 120\,e^-$. In the K' band, galaxies should be reliably separated from stars, *through the Galactic Plane*, with a loss of only 0.6 (1.2) magnitudes at $\ell = 90°$ (45°). However only few galaxies will be present up to these magnitudes, because of the strong dust extinction, even in K'. The corresponding losses in 92% reliability limiting magnitude, corrected for extinction, right through the Galactic Plane, are 1.4 and 3.6 magnitudes at $\ell = 90°$ and 45°, respectively, but only 1 magnitude or less for $|b| = 2°$ at these same longitudes.

Of course, the goal is not to have a highly reliable galaxy catalog everywhere, but instead a highly reliable galaxy catalog at high galactic latitudes and a moderately reliable (at the 50% level, see Table 2) galaxy catalog at low galactic latitudes. The results in Table 2 indicate that star/galaxy separation in the K' band will have at least 50% reliability everywhere, except in a narrow region extending from $\ell = 0°$, $b = 3°$ to $\ell = 45°$ (and $\ell = 315°$), $b = 0°$, and roughly 92% reliability at the Galactic Poles. It remains to be seen how extinction affects the detection of low galactic latitude galaxies.

6.   Prospects

In practice, the DENIS team will continue algorithmic development using simulated images, for the simple reason that we know everything we put in these images. In a second stage, we will take a dozen or so exposures of the same strips

Table 2.   Reliability Limits for Star/Elliptical Galaxy Separation

ℓ	$|b|$	Reliability	K'	K^0	J	K^0	I	K^0
90°		92%	13.2	13.2	14.4	13.4	16.5	14.7
90°		50%	14.2	14.2	14.7	13.7	18.0	16.2
0°	10°	92%	12.9	12.8	14.3	13.1	15.9	13.9
0°	10°	50%	13.4	13.3	14.5	13.3	16.7	14.7
0°	5°	92%	12.8	12.6	14.3	12.8	15.1	12.6
0°	5°	50%	13.6	13.4	15.3	13.8	16.9	14.4
45°	2°	92%	12.6	12.2	14.5	12.4	14.9	11.3
45°	2°	50%	14.1	13.7	16.2	14.1	16.7	13.1
45°	1°	92%	12.5	11.6	14.4	11.2	14.6	9.2
45°	1°	50%	13.9	13.0	15.5	12.3	16.7	11.3
45°	0°	92%	12.0	9.6	14.0	7.0	14.5	3.6
45°	0°	50%	13.2	10.8	15.4	8.4	16.8	5.9
90°	2°	92%	12.6	12.3	14.4	12.7	15.0	12.2
90°	2°	50%	14.0	13.7	15.8	14.1	17.4	14.6
90°	1°	92%	12.6	12.2	14.3	12.3	14.5	11.0
90°	1°	50%	13.8	13.4	16.1	14.1	16.7	13.2
90°	0°	92%	12.6	11.8	14.0	10.9	14.5	9.3
90°	0°	50%	13.8	13.0	15.7	12.6	16.6	11.4
180°	0°	92%	12.7	12.3	14.4	12.4	14.5	11.2
180°	0°	50%	13.5	13.1	15.6	13.6	17.0	13.7

K^0 is the equivalent K *galaxy* magnitude corrected for extinction.
The accuracy on the magnitude limits is $\Delta m \simeq 0.2$.

of sky, and extract the galaxies on the co-added strips, using the algorithms and parameters obtained with the image simulations. We will then iterate on the extraction parameters analyzing the individual strips and comparing to the master list obtained with the co-added strip. In this way, we can optimize the algorithm parameters, no longer being subject to the oversimplified galaxy models used in the image simulations. In addition, this will return us the *selection function*, which we need to know as a function of band, object type and magnitude, star field density, extinction, background (instrumental + sky) and PSF. The accurate determination of the selection function will require a substantial number of strips to be observed a dozen times or so.

To save time, we are considering (following a suggestion by D. Lynden-Bell) focusing on the Galactic Pole, and mimicking the effects of confusion at low galactic latitudes by superposing low galactic latitude fields on top of polar fields, and then attempting to reextract the galaxies that we had indeed detected in the polar field. We would then have to add photon noise to the background to incorporate the effects of extinction and lower surface brightness galaxies at very low galactic latitudes.

References

Bahcall, J. N., & Soneira, R. M. 1980, ApJS, 44, 73

Burstein, D., & Heiles, C. 1982, AJ, 87, 1165

Harmon, R. T., & Mamon, G. A. 1993, in Sky Surveys: Protostars to Protogalaxies, B. T. Soifer, San Francisco: A.S.P., 15

Jablonka, P., & Arimoto, N. 1992, A&A, 255, 63

Mamon, G. A. 1994, in Near-Infrared Sky Surveys, A. Omont, Dordrecht: Kluwer, in press

Mamon, G. A., & Contensou, M. 1993, DENIS internal report

Mamon, G. A., & Soneira, R. M. 1982, ApJ, 255, 181

Meurs, E. J. A., & Harmon, R. T. 1988, A&A, 206, 53

Wainscoat, R. J., Cohen, M., Volk, K., Walker, H. J., & Schwartz, D. E. 1992, ApJS, 83, 111

Young, P. J. 1976, AJ, 81, 807

Discussion

N. Epchtein: Status of the DENIS project (as of January 1994). The DENIS test operations just started at La Silla in December 1993. The focal instrument, built under the responsibility of Paris Observatory and provisionally equipped with 2 NICMOS3 infrared cameras (J and K_{short}), has been running at the ESO 1 meter telescope for approximately one month. The first sets of images are currently being analyzed in our data analysis centers at Paris (IAP) and Leiden Observatory. Several strips of 30° in declination and 12' in RA have been acquired, showing that the survey strategy that we have defined works properly. The limiting magnitude of $K_{short}=14$ (3σ in 10 sec integration time) is reached and will probably be improved after optimization of the all instrument. The I band camera (a 1024^2 Tektronix CCD) will be set up in March 1994, and we expect to start the survey operations by October 1994 when the telescope will be dedicated full time to the project. We expect to complete the all southern sky coverage in less than 4 years.

J. Huchra: One important activity that you should engage in is thruth testing by obtaining images that are much deeper and have higher spatial resolution. In all these surveys ascertaining the completeness and uniformity limits are extremely important.

G. Mamon: I agree! In fact, we have applied for time on the ESO 2.2m telescope using the IRAC2 detector (0."5/pixel) to observe fields near the Galactic Plane. I do not know if ESO will give time to this project where the P.I. (me) is a theorist!

Unveiling Large-Scale Structures Behind the Milky Way
ASP Conference Series, Vol. 67, 1994.
C. Balkowski and R. C. Kraan-Korteweg (eds.)

How ESO-LV did and did not (?) look through the Galaxy

EDWIN A. VALENTIJN

Laboratory for Space Research, SRON-Groningen,
P.O.Box 800, NL-9700 AV Groningen, The Netherlands

Abstract. Recent results on the opacity of external galaxies are compared to results of extinction studies of the Galaxy. If our local environment has a similar opacity as found for ESO-LV Sb/Sc's, we would expect a total face-on optical depth $\tau \sim 1.3$. Galactic extinction maps and galaxy counts suggest an extinction towards the Southern Galactic hemisphere which is larger than in the Northern direction. In total, reddening due to cirrus could account for about $\tau = 0.5$. If one insists on the Galaxy to conform to ESO-LV Sb/Sc's, then cold molecular clouds should obscure a significant part of the Southern sky.

1. Introduction

Recent discussions about the degree of transparancy of the outer regions of Sb and Sc galaxies were triggered by the studies of Disney *et al.*(1989) and Valentijn (1990, hereafter V90). Here, we focus on the regions in the radial range between one exponential scale length and the outer regions of spiral galaxies, say at the 25th or the 26th blue isophote and discuss how results of external galaxies could possibly be matched with data of the Galaxy.

Often, the question arises: 'if galaxies like our Galaxy behave opaque, how is it then possible that we can observe from Earth a significant part of the Universe?'. Here, we address this question, along lines which were originally presented in Valentijn 1991 (hereafter V91).

The resolution of this issue is still speculative, but I want to present some of our findings to the participants of this conference as the results might help to resolve a number of poorly understood extinction properties of the Galaxy. One outstanding problem again showed up at this conference in the presentation by B. Burton: 'on one hand the Leiden/Dwingeloo Survey of HI shows a tight correlation between HI column density and IRAS 100 μm flux on small scales, while on the other hand, at large radii, the radial 100 μm flux density falls off much quicker than the HI column density'. This is perhaps one of the most important questions, which needs a better physical understanding.

The Galactic HI survey might learn us about the radial variation of gas to dust ratios and what we might expect in external galaxies – vice versa, the statistical extinction studies of external galaxies, might learn us what to expect for the radial variation of star to dust ratios in the Galaxy. Here, we address the latter question and in order to quickly show the various steps that are undertaken in the statistical evaluations I have compiled Figure 1. This graph shows the

The Opacity Study Ladder

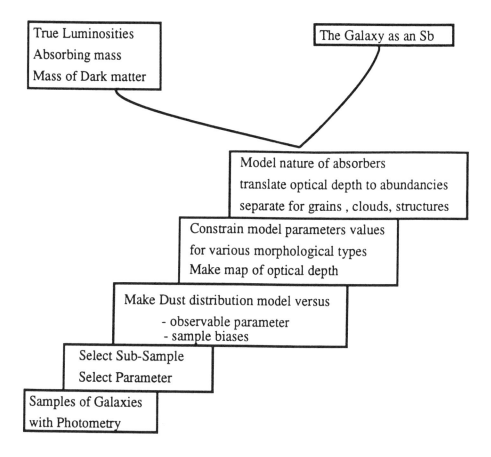

Figure 1.

great number of steps before arriving at a model for the Galaxy, and even when the last step is not correct, for instance in the case that our local environment in the Galaxy is different from the average Sb-Sc spiral, then this does not say much about the validity of the previous steps.

2. The Opacity Study Ladder

Figure 1 summarizes the various steps that are used in statistical evaluations of the mean opacity of the disks of spirals. After obtaining a sample of galaxies with detailed surface photometry (ESO-LV in our work) both a parameter and a sample have to be selected. It is then investigated how, observationally, the parameter value is related to the inclination angle of the galaxy on the sky.

Already at the very first step of the process (obtaining accurate photometry) the first debate starts. Huizinga and van Albada (1992, see also Huizinga's thesis, 1994) deduce transparent outer regions by assuming that the original visual diameter estimate of Lauberts was incorrect and that the ESO-LV value of the diameter at the 25th isophote (D_{25}) is correct, in spite of the notion by the authors of the ESO-LV catalogue that due to a technicality an artificial inclination dependency was introduced in the D_{25} listed in ESO-LV. Valentijn (1994, hereafter V94) presents a recipe for a small correction to D_{25}, which is confirmed photometrically by Peletier et al. (1994b) and numerically by Rönback and Bergvall (1994). After applying this recipe, the corrected ESO-LV D_{25} values are in a very good agreement with the original visual estimates for all inclinations. Both diameter values do not show any significant correlation with inclination and thus conform the expectations for optically thick outer regions. Unfortunately, the conclusions of Huizinga and van Albada are caused by neither using the original visual diameters, nor the corrected ESO-LV values.

Both Burstein et al. (1991) and V94 extensively discussed the problems with various sample/parameter combinations, the second step on the ladder. Burstein et al. rightfully warn against combinations which use as a parameter the apparent diameter or the apparent magnitude and for the sample selection a criterion which is also distant dependent. When redshifts are available such problems can be circumvented and, doing so, Burstein et al. conclude high optical thickness in the outer regions of spiral galaxies, confirming the V90 results, but objecting against his method. However, these objections are not relevant as in V90 the, distant independent, local surface brightness was used as the parameter, which does not suffer from the suspected ambiguities. This is further illustrated in V94, in which for a variety of sample selections, the same result is obtained: a very modest inclination dependency of the local surface brightness of Sc and Sc galaxies, at least at their effective radius.

When redshifts of the samples are known, indeed a number of problems can be circumvented, see also the interesting approach presented by Bottinelli et al. at this conference. Another way to trace distant dependent selection effects is to determine V/V_{max} values of the sample, while varying the cutoff limit of the sample. This way, V94 demonstrated that incomplete samples, with $V/V_{max} < 0.5$, exhibit an increase of isophotal diameters with increasing inclination angle, a property of transparent outer regions. On the other hand, complete samples with $V/V_{max} = 0.5$, which have been cut at a higher limit, do not show any significant increase of isophotal diameters with inclination angle, now conform the expectations for optically thick outer regions. I conclude, that the current literature of Sb,Sc galaxies still strongly supports the view that both their Blue isophotal diameters and local surface brightness hardly depend on inclination. For a simple slab model with a homogeneous mix of stars and absorbers, or for a sandwich model with a more transparent outer region, the hitherto determined regressions (values of C) correspond to optical depth listed in Table 1.

Table 1.   Very global face-on B band opacity values of Sb galaxies

Center	$C \sim 0$	$\tau \sim 5\text{-}30$
effective radius	$C \sim .2$	$\tau \sim 1\text{-}2$
D_{25}, D_{26}	$C \sim .2\text{-}.4$	$\tau \sim 1$

Note, that at this state of the process (fourth box on the ladder, Figure 1) we need not yet to define the nature of the absorbers, we only assume them to be well mixed with the stars.

However, as outlined in V90, the tabulated radial variation of τ is approximately a factor 10 less than the radial variation of the emitted star light. Thus, the data suggest an absorbing component with an exponential scale length larger than that of the stars (this was actually the motivation for the alternative two-disk rotation curve solution presented by González and Valentijn, 1991). In a recent study, comparing the near infra-red K band scale length with that in the Blue band, for a representative sample of 38 spiral galaxies, Peletier et al. (1994a,b) find indeed that the ratio (B/K) is in the range 1.3-1.9. A detailed application of a radial version of the equation of radiation transfer resulted in detailed spatial models for the absorbing components. The observed B/K scale length ratios, together with the noted absence of a diameter increase with inclination, support a two component model for the absorbing material: one component which follows the distribution of the stars (as indicated by the K photometry) and one component which ought to have at least a two times larger scale length than that of the stars.

The dust component that follows the stars, is nearly transparent at the effective radius and its spatial distribution matches to the dust as detected by IRAS. Note, that Bothun and Rogers (1992) also claim the IRAS detected dust to be essentially transparent at the visual bands.

The second component, with a scale length larger than that of the stars, ought to have a dust temperature $< 20K$ and escaped from detection by IRAS. The inclination studies for simple sandwich models indicate at the effective radius a value of $\tau \sim 1.3$, which corresponds to $A_B \sim 0.6$. In V90 it was proposed that cold molecular clouds could largely contribute to absorption by this second component.

So if, after all, we wish to evaluate whether the opacity studies of external galaxies could match to the local properties of the Galaxy, we have to evaluate the question whether a local value of $\tau = 1.3$ is appropriate, taking into account that it should contain one component with a scale length equal to that of the stars (with $\tau = 0.3\text{-}0.4$) and one component associated with cold dust with a scale length much larger than that of the stars (with τ around 0.9).

3. The Galaxy as an Sbc with $\tau = 1.3$

Of course, our Galaxy could be special, or our local environment could be special, allowing for better local transparency than suggested by the inclination tests of external galaxies. But, it seems unsatisfactory to close the issue with such a statement. Let us now explore the possibilities that actually the local Galaxy behaves like typical ESO-LV disks.

The ESO-LV data base itself contains valuable information about the extinction in the Galaxy, which data have the unique property that they have been acquired in a very homogeneous fashion over a large part of the Southern Galactic Hemisphere and for a significant part of the Northern Galactic Hemisphere. (See V91 for a preliminary report on this Section).

By applying statistical techniques to measured surface brightness values of about 12000 ESO-LV galaxies Choloniewski and Valentijn (1991, hereafter CV91) have obtained a Galactic extinction map. The map corresponds to the results of an alternative, more conventional approach, using the B-R reddening. The map also closely corresponds to the IRAS map obtained at the South Pole by Boulanger *et al.* Remarkably, however, the map suggests a much higher extinction towards the South Pole than towards the North. Given the homogeneity of the data acquisition, it is difficult to see how this could have entered our data in an artificial way. Thus, the data suggest that the Sun is located slightly above the main absorbing (cirrus) layer of the Galaxy and actually shows $A_B = 0.2-0.3$ towards the South Pole. In Table 2 these 'reddening' values are listed together with the corresponding values of τ for a homogeneous slab model.

Table 2.   The local Galaxy with total $\tau = 1.3$

	A_B		τ	
	North	South	North	South
Reddening (CV 1991)	0.05	.2 - .3	0.1	0.4
Molecular clouds	0.15	.25	0.3	0.5

Fig 2. shows the count rate (number of galaxies per square degree) of ESO-LV galaxies, as a function of Galactic latitude. We have made many of such plots, for instance excluding Abell cluster galaxies, or poor group galaxies, taking different cross cuts through the sky (separated in RA), but it is always found that there is a higher count rate in the North than in the South.

Fig 3. presents the differential count rate normalized with a Euclidian world model. Evidently, the count rate is a factor 1.6 higher in the North than in the South, which value is reproduced when plotting the count rate of the faint background galaxies, which have been recorded on the ESO-LV images.

The count rate difference North/South is NOT correlated with the reddening difference North/South obtained by CV91, and thus the variation in count rate can not be caused by the component that is responsible for the reddening (see V91 Fig 5)!

It is possible, that we are observing the effect of an intrinsic overdensity of galaxies in the Northern Hemisphere, essentially due to the Hydra and Centaurus clusters, but this seems not directly to match to the repetition of the count rate

E.A. Valentijn

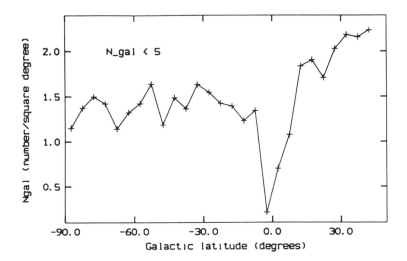

Figure 2.   The count rate (number of galaxies per square degree) of
ESO-LV galaxies as a function of Galactic latitude, selecting galaxies
not identified with clusters or rich groups of galaxies ($N_{gal} < 5$).

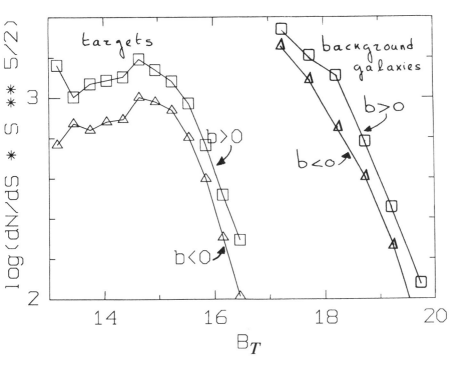

Figure 3.   Differential counts normalized with a Euclidian world
model of the ESO-LV galaxies and background galaxies found in the
ESO-LV survey (in bins of 0.3^m total B magnitude). Squares for the
Northern Galactic hemisphere, triangles for the Southern hemisphere.

difference at faint magnitudes (18-20) or when neglecting rich cluster galaxies.
This item deserves a more extensive investigation by extending galaxy counts
further towards the North.

It is interesting, but speculative, to assign the count rate difference to the
effects of molecular clouds, as suggested by the inclination studies. In order
to create a count difference North-South of 1.6 in a disk with total $\tau = 1.3$
(with 0.5 being caused by cirrus type of dust) the presence of molecular clouds
in the Northern hemisphere should result into an effective $\tau = 0.3$, while in
the Southern hemisphere this value is then 0.5. Numerical experiments with
randomly distributed clouds indicate that a value of $\tau = 0.5$ would imply for
an observer located in the disk that about 40-50% of the sky will be obscured
at Galactic latitudes 20-40°, and 20% of the sky will be obscured at latitudes
> 40°. However, these numbers are subject to strong statistical fluctuations,
(e.g. one nearby cloud could obscure half of the sky).

At this conference, I wanted to present these puzzling results, as your comments might help progress in this complicated field. Indeed there is an answer
to the question: 'if galaxies like our Galaxy behave opaque, how is it then possible that we can observe from Earth a significant part of the Universe?'. This

answer is tabulated in Table 2. Also the spatial distribution of the reddening component (IRAS seen) and the cold component (larger than that of the stars) seems to obey the results of extinctions studies of external galaxies. But the implication would be that we are missing a significant fraction of galaxies in our surveys, without noticing it. That is not an attractive conclusion.

References

Burstein D., Haynes M.P, Faber S.M., 1991, Nature **353**, 515

Burton B., this proceedings

Bothun G.D., Rogers C., 1992 AJ **103**, 1484

Bottinelli *et al.*, this proceedings

Choloniewski J., 1991, MNRAS **250**, 486

Choloniewski J., Valentijn E.A., 1991, (=CV91) ESO Messenger, **63**, 1

Disney M., Davies J.I., Phillipps S., 1989, MNRAS **239**, 939

González-Serrano J.I., Valentijn E.A., 1991, (=GV91) A&A **242**, 334

Huizinga J.E., van Albada T.S., 1992, MNRAS **254**, 677

Lauberts A., 1982, The ESO/Uppsala Survey of the ESO (B) Atlas, ESO, München

Lauberts A., Valentijn E.A., 1989, The Surface Photometry Catalogue of the ESO-Uppsala Galaxies (=ESO-LV), ESO, München

Peletier R.F., Valentijn E.A., Freudling W., and Moorwood A.F.M. 1994 Infrared Astronomy with arrays, p 149, Ed. I. Mc Lean, Kluwer

Peletier R.F., Valentijn E.A., Moorwood A.F.M., and Freudling W., 1994, A&A Suppl. submitted

Rönnback J., Bergvall N., 1994, A&A in press

Valentijn E.A., 1990, (=V90) Nature **346**, 153

Valentijn E.A., 1991a, (=V91) Proc. IAU Symp. No 144, ed. H.Bloemen, Kluwer, p 245

Valentijn E.A., 1991b, ESO Messenger **63**, 45

Valentijn E.A., 1994, (=V94) MNRAS **266**, 614

Discussion

D. Lynden-Bell: Have you correlated neutral hydrogen with the best ESO maps? This would be very useful for doing studies in the plane where the Hartmann and Burton nice 21-cm results are all there is!

E. Valentijn: No but we plan to.

H. van Woerden: Observations of 21-cm line profiles do not directly give HI column densities. They give profile integrals $\int T_b \, dV$; the conversion to N(HI) depends on the distribution of the spin temperature T_S along the line of sight. If $T_S \gg T_b$ everywhere, i.e. for low brightness temperatures, there is no problem : N(HI) is to good accuracy proportional to $\int T_b \, dV$. But at low latitudes, high T_b values, probably comparable to T_S, are observed ; the optical depths may become high, and N(HI) values based on the low-τ approximation are underestimates. Cold clouds may contain lots of HI seen in absorption (i.e. not contributing to the profile integral) or going unnoticed.

E. Valentijn: One has to use $\int T_b \, dV$ because it is all one has. Perhaps a non-linear correlation would be better?

G. Mamon: From the difference in obscuration between the NGP and SGP fields that you just presented, what do you derive for the distance of the Sun to the Galactic Plane?

E. Valentijn: From reddening studies we deduce a position of the Sun slightly above the plane of the dust layer (actual value not specified). However, from galaxy count studies, modelling the count difference N-S by compact clouds, we obtained a value for the position of the Sun of $\sim$ 12.5 pc above the plane, with a total thickness of the cloud layer of $\sim$ 95 pc. Based on Pioneer 10 data, Toller (1981) reports a position of the Sun 12.2 $\pm$ 2.1 pc above the plane.

J. Huchra: Will not any tests you make using α_K / α_B depend on the intrinsic stellar population gradients in the galaxies?

E. Valentijn: Yes, the question is important to address. The α_K / α_B ratios of 1.4-1.9 of course mean strong colour gradients, in fact from B-K $\sim$ 6.0 in the centers to B-K $\sim$ 3.5 at the outskirts. We evaluated these gradients in terms of radial population differences and come to the conclusion that they are too large to account for by 'standard' population variations.

L. Gouguenheim: I am impressed by your detailed study of selection effects. Nevertheless, I understand that the controversy is not closed ... To overcome these problems, we (Bottinelli, Gouguenheim, Paturel and Teerikorpi) used a

E.A. Valentijn

different approach. We used Tully-Fisher relations between either linear diameter or absolute magnitude and maximum rotational velocity together with kinematic distances to derive fiducial apparent diameters and magnitudes. Their comparison with observed quantities, as a function of disk inclination, lead to the conclusion that Sb to Scd galaxies are optically thick at the 25 magnitude arcsec^{-2} level. This study will be presented in my talk.

N. Lu: There have been a number of papers describing direct measurements of the opacity of spiral disks, however, the results are still controversial. Do you have any comments on this and any suggestion for further investigations?

E. Valentijn: I have reviewed several of these papers in the 1994 paper (MNRAS). For further investigations I regard a comparison of K and B photometry as the most promising.

Unveiling Large-Scale Structures Behind the Milky Way
ASP Conference Series, Vol. 67, 1994.
C. Balkowski and R. C. Kraan-Korteweg (eds.)

X-ray observations in the Zone of Avoidance

A.C. Fabian

Institute of Astronomy, Madingley Road, Cambridge CB3 0HA, UK

Abstract. X-ray observations offer an interesting window through the Zone of Avoidance. Soft X-ray absorption and source confusion cause some problems but future harder X-ray images should give a clear picture of distant clusters and quasars in the Galactic Plane. Several interesting clusters discussed here lie within 10 deg of the Plane are already well-studied in X-rays.

1. Introduction

The galaxy is mostly transparent to X-rays above a few keV, so the X-ray band is potentially an excellent window with which to observe the Zone of Avoidance (ZoA). However, source confusion, photoelectric absorption by our interstellar medium, Galactic Ridge emission cause serious problems to current studies, especially since the deepest present all-Sky surveys (ROSAT) are in the soft X-ray band below 2 keV where these problems are worst. There are however older surveys (eg HEAO-1) which are in the 2–10 keV band and are secure down to a flux level of $\sim 4 \times 10^{-11}\,\mathrm{erg\,cm^{-2}\,s^{-1}}$ over most of the Sky.

In this brief review I concentrate on the all-Sky distribution of galaxy clusters and point out that there are some very interesting clusters at low galactic latitude. Indeed the most X-ray luminous cluster within redshift $z = 0.2$ and the most luminous radio galaxy within $z = 0.5$ both lie within 7 degrees of the Galactic Plane!

2. Surveys, Source Confusion and all that

Many X-ray binaries, active stars, cataclysmic variables and RS Cvn systems along the Galactic Plane make it bright in all but the softest X-rays (when absorption overwhelms even nearby sources). A good image of the Plane in 2–10 keV X-rays is given by the EXOSAT scan presented by Warwick et al (1985). The instrument used for this was a broad-beam one (45 arcmin FWHM) and so does not resolve fine details. The Einstein Observatory Galactic Plane Survey of Hertz & Grindlay (1984) shows what can be done with an imaging telescope and discusses predictions for fractions of various classes of source. The ROSAT All-Sky Survey should eventually provide detailed maps of the Plane below 2 keV.

A diffuse ridge of hard emission was seen with early satellite instruments (Warwick et al 1979; Iwan et al 1982) and has now been well-mapped with Ginga (Koyama et al 1989). This last work has shown that much of the emission is

A.C. Fabian

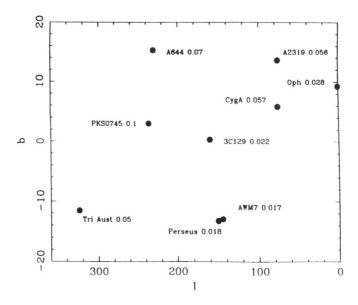

Figure 1.   X-ray bright clusters near the Zone of Avoidance, from
Lahav et al (1989) and Edge et al (1990). The redshift of the cluster
is indicated after the name.

thermal (a strong iron emission line at 6.7 keV is seen – see Yamauchi 1992 for
details of its distribution). The origin of this emission is not yet clear.

All-Sky surveys in the 2–10 keV band have been carried out with Uhuru,
Ariel V and HEAO-1.

3.   The distribution of clusters

A flux-limited sample $(S(2 - 10\,\text{keV}) > 1.7 \times 10^{-11}\,\text{erg cm}^{-2}\,\text{s}^{-1})$ of 54 clusters, complete above $3 \times 10^{-11}\,\text{erg cm}^{-2}\,\text{s}^{-1}$, shows several clusters close to the
Galactic Plane (Lahav et al 1989; Edge et al 1991). 9 out of 34 clusters with
$S > 4 \times 10^{-11}\,\text{erg cm}^{-2}\,\text{s}^{-1}$ have $|b| < 20°$ and 4 lie within $|b| < 10°$ (Fig. 1).
These are the Ophiuchus cluster, Cygnus A, 3C 129 and PKS 0745-191. All but
3C 129 have been well studied and imaged with EXOSAT and ASCA. Cygnus
A, 3C 129 and PKS 0745 were of course known earlier from their radio sources,
the Ophiuchus cluster (the second X-ray brightest cluster in the whole Sky) was
independently discovered by Johnston et al (1981) from X-ray data and Wakamatsu & Malkan (1980) from Schmidt plates (see contribution by Wakamatsu .

The X-ray luminosity of clusters varies approximately as the cluster mass
cubed, so it is a sensitive measure of the most massive bound objects known. The
most luminous clusters mark out the 'mountain peaks' of the galaxy distribution.
In this way the Ophiuchus cluster is to Hawaii as the Shapley Supercluster is
to the Himalayas! X-ray spectra of this cluster have been shown by Arnaud
et al 1987 and Yamashita et al (1992). The Ophiuchus cluster appears isolated
(despite the claim by Djorgovski et al 1990). It will be interesting to see whether

it is connected through large-scale structure to 3C129, or whether 3C129 is connected with the Perseus clsuter and AWM7 (see Fig. 1).
PKS 0745-191 is the most luminous cluster of the brightest 54 and is the most luminous within $z = 0.2$, with $L_X(2 - 10\,\mathrm{keV}) = 3 \times 10^{45}\,\mathrm{erg\,s^{-1}}$. It has a massive cooling flow (Fabian et al 1985; Arnaud et al 1987). Cygnus A is the most luminous radio galaxy within the low redshift Universe and lies in an X-ray luminous cluster (Arnaud et al 1984; 1987).
Work carried out with O. Lahav, looking at the HEAO-1 A2 database with K. Jahoda and the RASS with H. Böhringer have failed so far to clearly show any emission from the very nearby Puppis cluster (see contribution by Lahav).

4. The Future

The RASS should reveal many extragalactic objects within $|b| = 20°$, despite the soft band. ASCA pointings along the Galactic Plane should serendipitously reveal extragalactic sources. The GIS instrument has a field of view of 50 arcmin and is sensitive to above 10 keV; images of Galactic sources for example can be searched for distant clusters and AGN.
Later missions yet to be launched will gradually open up the ZoA for X-ray studies. So far however there is no approved mission to carry out a deep imaging survey above 2 keV. That would transform our knowledge of this part of the Sky. Many more extreme objects may lie hidden!

References

Arnaud, K.A., Fabian, A.C., Eales, S.A., Jones, C., Forman, W., 1984. MNRAS, 211, 981

Arnaud, K.A., Johnstone, R.M., Fabian, A.C., Crawford, C.S., Nulsen, P.E.J., Shafer, R.A., Mushotzky, R.F., 1987. MNRAS, 227, 241

Djorgovski, S., Thompson, D.J., De Carvalho, Mould, J.R., 1990. AJ, 100, 599

Edge, A.C., Stewart, G.C., Fabian, A.C., Arnaud, K.A., 1990. MNRAS, 245, 559

Fabian, A.C., et al. 1985. MNRAS, 216, 923

Hertz, P., Grindlay, J.E., 1984. ApJ, 278, 137

Iwan, D., et al. 1982. ApJ, 260, 111

Johnston, M.D., et al. 1981. ApJ, 245, 799

Koyama, K. et al. 1989. Nature, 339, 603

Lahav, O. Edge, A.C., Fabian, A.C. & Putney, A. 1989, MNRAS, 238, 881

Wakamatsu, K. & Malkan, M. 1981, PASJ, 33, 57

Warwick, R.S., Pye, J.P., Fabian, A.C., 1980. MNRAS, 190, 243

Warwick, R.S., et al. 1985. Nature, 317, 218

Yamashita, K., in Frontiers of X-ray Astronomy, eds Tanaka Y., Koyama, K., Universal Academy Press, Tokyo

Yamauchi, S., 1992. PhD thesis, ISAS, Tokyo

Discussion

J. Huchra: There has been a proposal for a higher energy (2-10 keV) X-ray survey telescope by Burg, Giacconi and others. This beast is called the WFXT or wide field X-ray telescope and will image with enough spectral resolution to both find clusters and measure their redshifts at the same time. This was proposed for the last round of NASA SMEX missions but did not make it. We are still trying for funding from other sources.

Unveiling Large-Scale Structures Behind the Milky Way
ASP Conference Series, Vol. 67, 1994.
C. Balkowski and R. C. Kraan-Korteweg (eds.)

The Expansion and Update of the "Southern Redshift Catalogue"

A.P. Fairall

Department of Astronomy, University of Cape Town, Rondebosch, 7700 South Africa

Abstract. A major expansion of this catalogue is being undertaken. For each galaxy, individual heliocentric velocity measurements, errors and appropriate flags are now listed, as well as the existing 'best estimate'. Sources are immediately identified by author and abbreviated reference. The new version will be completed in two segments; the first, that covering R.A. 8 to 16 hours (ie. that affected by the southern Milky Way) was made available - via Internet - at the time of the conference.

1. Background

The present catalogue was started in 1980. Over the years, it has grown and been updated in an attempt to include all published redshift measurements (optical and radio) for galaxies south of the Equator with $cz < 75000$ km s^{-1}. The last published version of the catalogue is Fairall and Jones (1991).

It is one of a number of similar compilations. All-sky redshift catalogues include the Harvard-Smithsonian ZCAT (organised by J. Huchra and currently maintained by C. Clemens), the French LEDA database (headed by G. Paturel), and the IPAC database at Caltech. A collaborative spirit exists within this community and numerous cross checks have been carried out to improve completeness and to identify errors.

The author was also a collaborator for the intended updated version of the 'Catalogue of Radial Velocities of Galaxies' (original version by Palumbo et al 1983). That compilation filled a definitive role, since it was the only one to list all original individual measurements and sources. Regrettably, various circumstances have prevented the completion of this catalogue. Thus it seemed appropriate to modify and expand the format of the Southern Redshift Catalogue so to fill the vacuum. In doing so, there would also be less duplication of the work being put into ZCAT etc.

2. The New Version

More than sixty new references (up to mid 1993) have been added, and the process is ongoing. Format expansion (including recovery of original measurements and errors) and update has so far been completed for the segment R.A. 8 to 16 hours, which has nearly 9000 redshift measurements for over 5000 galaxies.

The remaining segment (in preparation 1994) is expected to have an estimated 12000 measurements for 10000 further galaxies.

Aside from its existence as a database, the catalogue plays an important role. Not only does it identify galaxies already observed (and for which further valuable telescope time should not be invested), but it also identifies the inevitable discrepant redshifts (about 1-2% of published redshifts are in error).

Unfortunately, the increased use of fibre spectrographs, with the associated burden of reducing their abundant data so to produce carefully checked measurements, has led to large numbers of redshift measurements as yet unpublished, and not included in the present work.

3. Sample of format

```
E494-G14
  8 02 41 -23 50 59      8 00 33 -23 42 36   242.2    3.7
 (5650) 50 R             R148 Kraan-Korteweg Huchtmeier 1992 AA
266,150


F-1144/E369-N03
  8 02 30 -34 30 08      8 00 36 -34 21 45   251.2   -2.0
 3310 100 O       *      R090 Fairall 1988 MNRAS 233,691


  8 01 08 -66 09 15      8 00 44 -66 00 51   279.0 -18.0
 12324   36 O     *      R261 Strauss et al 1992 ApJSupp 83,29


  8 01 50 -72 24 21      8 02 16 -72 15 51   285.0 -20.7
 13725   40 O     *      R261 Strauss et al 1992 ApJSupp 83,29
 13721   35 O EM  *      R263 Sekiguchi Wolstencroft 1992 MNRAS
255,581
 13733   81 O EM  *      R263 Sekiguchi Wolstencroft 1992 MNRAS
255,581
                        NB Last two above N and S components
 13726   49 O EM  *      Weighted mean


PKS
  8 05 16  -1 11 07      8 02 43  -1 02 36   222.7   15.8
 26320      O     *      C511 Sargent 1975 in Mem RAS 79,75.


  8 04 06 -53 38 58      8 02 50 -53 30 26   267.8 -11.7
 18111   40 O     *      R261 Strauss et al 1992 ApJSupp 83,29
 18200      O     *      R268 Staveley-Smith et al 1992 MNRAS
258,725
 18129   54 O     *      Weighted mean
```

4. Availability

The catalogue is arranged in separate files for each hour of Right Ascension, as well as reference lists etc. It can be accessed over Internet at uctvax.uct.ac.za with username "redshift" and password "galaxy".

References

Fairall, A.P. & Jones, A., 1991. Publ. Dept. Astr. Univ. Cape Town, No. 11.

Palumbo, G.C.C., Tanzelli-Nitti, G., & Vettolani, G., 1983, Catalogue of Radial Velocities of Galaxies, Gordon and Breach.

Part III

Optical Searches on Sky Surveys

Optical Identification of Galaxies along the Northern Milky Way: a Status Report on a Project at Innsbruck

R. Seeberger, W. Saurer, R. Weinberger, G. Lercher

Institut für Astronomie der Universität Innsbruck, Technikerstraße 25, A-6020 Innsbruck, Austria

Abstract. In fall 1989 we have started an extensive optical survey for galaxies on Palomar Sky Survey red-sensitive prints that eventually comprised the entire Northern Galactic Plane; more than 5 000 galaxies were found. We report on several aspects of this project and on follow-up observations (H I, optical spectroscopy).

1. Introduction

In recent years, a marked upsurge of investigations in the optical, infrared, and radio range of the zone of avoidance (ZOA) can be noted. By weighing up the pros and cons of searches in the three wavelength regimes, (deep) optical identifications/observations would be advantageous, provided that the Galactic Plane is reasonably transparent. That this is the case, was demonstrated by, e.g., Weinberger (1980), Kraan-Korteweg (1989), and Saito et al. (1990, 1991), where - particularly along the Southern Galactic Plane - numerous galaxies have been optically discovered.

At present, our group in Innsbruck has just completed a survey of the entire Northern Galactic Plane and of a part of the Southern Galactic Plane on POSS red-sensitive prints, by searching the region $33° \leq \ell \leq 240°, -5° \leq b \leq +5°$, but extending from $b = -10°$ to $+10°$ in the longitude range $80° \leq \ell \leq 130°$ (see section 3).

A compilation of (hopefully) all published galaxies and quasars that were ever optically identified within $0° \leq \ell \leq 360°, -5° \leq b \leq +5°$ (see section 2) served as prelude to this project.

Our project is a resumption of similar, but much less extensive investigations in/of the ZOA at Innsbruck about 15 years ago (Weinberger et al. 1978; Weinberger 1980; Pfleiderer et al. 1981).

2. Compilation of published extragalactic objects in the ZOA

A literature search for published objects of certain types might, in the era of large machine-readable data compilations, be considered as obsolete. This would, however, be an entirely naive view, since large data banks of specific kinds of objects usually consist of compilations of more or less large published lists and catalogues but - with the possible exception of the most recent years - single objects, perhaps even "hidden" in papers devoted to different scientific aims, remain undetected for this purpose. Since it was our intention to strive for completeness (a goal that can, of course, hardly be reached), we have carried out a detailed, tedious literature search for published optically identified extragalactic objects, i.e., galaxies and quasars within $|b| \leq 5°$.

We started the detailed literature examinations from the year 1970 onwards, but included hopefully all the (few) catalogues or papers containing low-latitude extragalactic objects that were published before that year. Our search is based on the semi-annual volumes of *Astronomy and Astrophysics Abstracts* which are published for the Astronomisches Recheninstitut by Springer Verlag.

We made finds in altogether 67 papers, including papers that were published up to mid 1993, but added another three papers that were (or will be) published later. We examined each object individually except the galaxies in the huge lists of Saito et al. (1990, 1991), and Yamada et al. (1993) on the POSS, on Hoessel et al's (1979) infrared survey or on the ESO R / SERC J surveys. These examinations were done to check whether the objects are indeed galaxies, to determine their positions (better than $\pm 0.'3$), and to estimate their maximum diameters. We found, e.g., that a considerable number of "galaxies" turned out to be objects of other kinds, like reflection nebulae, plate flaws, or were not present at all within a couple of arcmin around the published positions; this turned out to be particularly true for MCG sources. The positions of many objects were inaccurate, often much worse than 1'.

The positions that we are able to present allow an unambiguous nomenclature of the ZOA galaxies by designating them via their galactic coordinates. In addition to the equatorial and galactic positions and the maximum optical diameters we tried to find the most detailed morphological type, took a great deal of care on being up-to-date with the radial velocities, and cross-checked the more than 2 000 galaxies with the IRAS Point Source Catalogue.

In Fig. 1 we show the distribution of all the objects, in Table 1 the first entries in our compilation, and in Fig. 2 a plot of the radial velocities.

3.   Optical Search

For our optical search we made use of POSS I E (red-sensitive) prints, since only few copies of the new POSS II atlas are available up to now. A microscope with 16 X magnification was chosen as survey instrument. The identification procedure was as follows: one person did the survey work and marked all objects that could possibly be galaxies; a second person was responsible for the final inclusion or rejection of these candidates using E prints (plus O prints plus infrared prints plus the (few) POSS II films available). The search resulted in more than 5 000 (accepted) galaxies.

As a second step, positions were determined with an accuracy of typically 10" or less. Then, diameter measurements were made by taking both the diameters of bulges (if visible) and (always) the maximum recognizable optical diameters on both the POSS E and O (blue-sensitive) prints.

In Fig. 3 we show the distribution of the galaxies, and in Fig. 4 a section displaying how the diameters can be graphically coded (the sizes representing their apparent (maximum) diameters). In Fig. 5 we present the same region as in Fig. 4, but use symbols the size of which increases with the number of

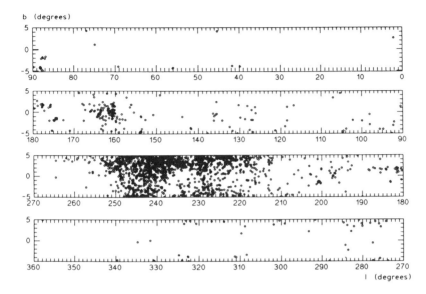

Figure 1. The distribution of published optically identified extragalactic objects along the entire Galactic Plane.

Table 1. A compilation of optically identified and published extragalactic objects within $\ell = 0° - 360°$; $|b| \leq 5°$

ZOAG	name	α(B1950.0)	δ(B1950.0)	α(J2000.0)	δ(J2000.0)	type	∅'	rad.vel.	ref.	IRAS
G118.44+1.25	WEIN9	00 08 00	+63 29.5	00 10 39	+63 46.2		0.2		41	00080+6329
G118.49−1.95	WEIN10	00 12 26	+60 20.0	00 15 06	+60 36.7	S	0.8	4512	41	
G118.97−3.31	IC10	00 17 40	+59 02.5	00 20 23	+59 19.1	IBm	5.5	−344	01	00175+5902
G121.33−4.24	MCG+10−02−01	00 36 17	+58 19.2	00 39 08	+58 35.7	S	1.0		02	00362+5819
G126.14−0.21	GT0116+622	01 16 01	+62 13.5	01 19 17	+62 29.2	—			62	
G126.77−4.63	MCG+10−03−01	01 17 15	+57 45.9	01 20 24	+58 01.6		0.4		02	01172+5746
G126.27+1.16	WEIN11	01 18 28	+63 34.6	01 21 48	+63 50.3		0.1		41	
G126.79−1.81	WEIN12	01 19 52	+60 33.6	01 23 07	+60 49.3		0.3		41	
G127.18−1.82	WEIN13	01 23 02	+60 40.5	01 26 18	+60 56.1		0.4		41	
G127.60−4.11	MCG+10−03−02	01 23 52	+58 11.0	01 27 05	+58 26.5		0.4		02	
G127.11+0.54		01 25 08	+62 51.0	01 28 30	+63 06.5		0.1	5490	38	
G128.70−1.61	WEIN15	01 35 28	+60 28.4	01 38 50	+60 43.6		0.6		41	
G128.15+1.43	WEIN14	01 35 30	+63 33.8	01 38 59	+63 49.0		0.2		41	
G130.53−4.92	CGCG559−02	01 44 16	+56 52.3	01 47 35	+57 07.2		0.3		15	
G129.86−1.26	WEIN16	01 45 18	+60 35.2	01 48 45	+60 50.1		0.3		41	
G131.81−4.28	5Zw153	01 54 30	+57 11.6	01 57 54	+57 26.2		0.6		02	
G130.29+1.65	WEIN17	01 54 41	+63 19.4	01 58 19	+63 34.0	S	0.6		41	01546+6319
G132.45−3.38	MCG+10−04−01	02 00 50	+57 54.1	02 04 18	+58 08.5		0.4		02	
G134.27−1.90	WEIN18	02 17 31	+58 45.5	02 21 07	+58 59.2		0.3		41	02175+5845
G136.27−1.91	WEIN20	02 31 40	+58 01.4	02 35 20	+58 14.5		0.2	5721	41	02317+5801
G135.83−0.57	Maffei1	02 32 36	+59 26.0	02 36 20	+59 39.0	E/S0?	2.0		15 17	
G135.56−0.34		02 33 21	+59 23.4	02 37 05	+59 36.4	E?	0.1		19	
G137.80−4.92		02 33 26	+54 39.3	02 36 59	+54 52.3	Sc	0.4		51	
G136.21−0.90	3C69	02 34 19	+58 58.9	02 38 03	+59 11.9	st	z=0.458	58	
G136.50−0.33	Maffei2	02 38 09	+59 23.4	02 41 55	+59 36.2	SBb(s)	5.5	−20	17	02381+5923

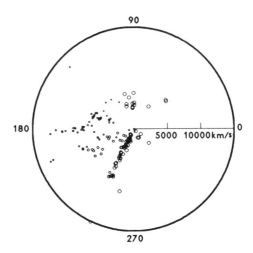

Figure 2.   One example of our data sets: V_{hel} of those galaxies having
$\leq 15\,000$ km/sec. Symbol sizes represent apparent optical (maximum)
diameters.

galaxies per square degree; such figures can, e.g., be used for the recognition of galaxy cluster candidates - several candidates of this kind are evident in Figs. 4 and 5.

In the following two sections we will describe how subsamples of the galaxies that we found were used for H I and optical observations. We also plan several observing runs for the (near) future in order to both investigate extragalactic aspects of a kind as discussed below and galactic aspects, like the determination of interstellar extinction along the sight to many galaxies with the aim to eventually obtain a detailed dust model of the Milky Way.

4.   H I Measurements of Galaxies in Selected Regions

As part of a program to carry out extensive H I observations of galaxies in selected regions of the Galactic Plane, 21cm observations in the NW region of the Pisces-Perseus-Supercluster were performed with the 100m radio telescope at Effelsberg, in cooperation with W.K. Huchtmeier, MPIfR Bonn. We observed about 150 galaxies and detected ca 1/3 of them; a few typical H I profiles are shown in Fig. 6.

The primary motive of these observations was to find out whether this well known supercluster is veiled by galactic extinction in the region $80° \leq \ell \leq 110°$, since by examining the distribution of the objects in optical galaxy catalogs we had gained this suspicion. Indeed, for 24 galaxies we found velocities in a range that fits well to the values usually accepted for the Pisces-Perseus-Supercluster, which could thus for the first time be pursued deep into the Galactic Plane, from the south up to a galactic latitude of $b \approx -5°$. This work was recently accepted by Astronomy and Astrophysics (Seeberger et al. 1994).

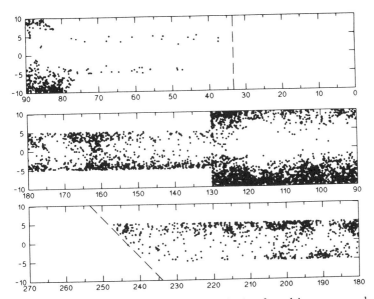

Figure 3.　The distribution of the galaxies found in our search.

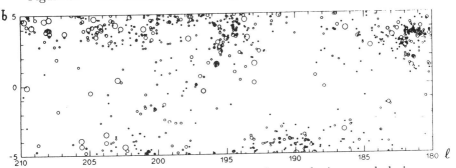

Figure 4.　The distribution of a part of our galaxies; symbol sizes represent apparent (maximum) optical diameters.

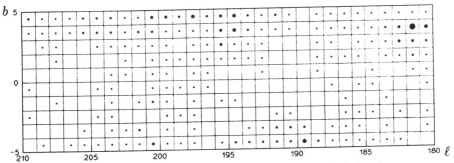

Figure 5.　Same region as in the previous figure; symbol sizes increase with the number of galaxies per square degree.

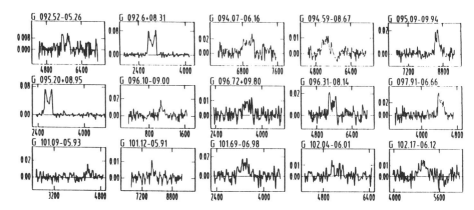

Figure 6.   H I profiles of a few of our detected galaxies. Fluxes (in Jy) are plotted versus heliocentric velocities (in km/sec).

In addition to this new extension of the PP-Supercluster we noticed, first from the distribution of known optical galaxies, another possibly very long structure that runs almost parallel to the southern border of the Galactic Plane; we observed the (easternmost?) part of it, extending up to $\ell \approx 110°, b \approx -5°$. Further observations are needed to check the reality and extent of this structure.

5. Optical Spectroscopy

Optical spectroscopy was carried out with the 1.92m telescope of the Observatoire de Haute Provence and the 1.82m telescope of the University of Padova at Cima Ekar. The aim of these investigations is twofold: firstly we try to investigate the three-dimensional structure of selected regions on the sky, secondly our catalogue provides a basis for the search for Active Galactic Nuclei. Unfortunately, due to repeatedly bad weather conditions in the past two years, we had less observations than we hoped for. Up to now we have observed about 50 galaxies with acceptable signal-to-noise ratios (several also outside the region optically searched).

Measurements of the radial velocities were made mainly in the region $80° \leq \ell \leq 110°$. With the chosen resolution we could span a wavelength-range from 3700 to 7700 Å and 1 pixel corresponds to ≈ 7 Å. The determination of the radial velocities was done by means of both absorption- and emission-lines (when visible). Because in most cases several lines $(5-10)$ could be used we estimate the accuracy to be ± 100 km/s. For several galaxies we had both optical and H I radial velocities and noted an excellent agreement between them.

The search for Active Galactic Nuclei is carried out in collaboration with P. Rafanelli (Univ. Padova). Because bright bulges are a part of the characteristics of these objects they can be expected to be overabundant in a sample with high extinction. First results brought out about 20 galaxies which show emission lines. Most of them are probably H II-region galaxies. However, two galaxies

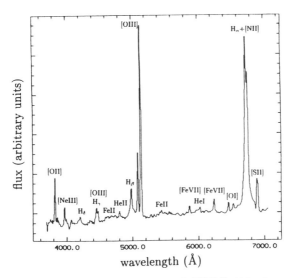

wavelength (Å)

Figure 7.   An optical spectrum of CGCG 493−2

could be identified as new Seyfert galaxies. As an example, the most prominent object (CGCG 493−2) is shown in Figure 7. This galaxy can be classified as Seyfert 1.5 with broad and narrow lines visible. In addition, from diagnostic line ratios an unusually high T_e (43 000 K) and a very low N_e could be derived.

Acknowledgments. This work has been supported by the "Fonds zur Förderung der wissenschaftlichen Forschung", Projekt Nr. P8325-PHY, and, a part of it, by the "Forschungsförderungsbeiträge der Universität Innsbruck".

References

Hoessel, J.G., Elias, J.H., Wade, R.A., and Huchra, J.P. 1979, PASP, 91, 41

Kraan-Korteweg, R.C. 1989, in: Reviews in Modern Astronomy 2, ed. G. Klare, Springer, Berlin, p. 119

Pfleiderer, J., Gruber, M.D., Gruber, G.M., and Velden, L. 1981, A&A, 102, L21

Seeberger, R., Huchtmeier, W.K., and Weinberger, R. 1994, A&A, in press

Saito, M., Ohtani, H., Asomuna, A., Kashikawa, N., Maki, T., Nishida, S., and Watanabe, T. 1990, PASJ, 42, 603

Saito, M., Ohtani, H., Baba, A., Hotta, N., Kameno, S., Kurosu, S., Nakada, K., and Takata, T. 1991, PASJ, 43, 449

Weinberger, R. 1980, A&AS, 40, 123

Weinberger, R., Elsässer, H., and Chini, R. 1978, Mitt. AG, 43, 116

Yamada, T., Takata, T., Djamaluddin, T., Tomita, A., Aoki, K., Takeda, A., and Saito, M. 1993, ApJS, 89, 57

Discussion

J. Huchra: In your work, did you use the IV-N Palomar Infrared Milky-Way Atlas?

R. Weinberger: Yes, but not for the search itself, but for classification of ambiguous cases.

H. van Woerden: In your optical search you plan to determine diameters. Can you estimate to what isophote these diameters will correspond?

R. Weinberger: According to Abell (1966), the surface brightness limiting magnitude of the POSS E (=red-sensitive) plates is about 25 mag arcsec^{-2} ; correspondingly, our diameters refer to about this isophote.

C. Balkowski: How did you select your sample of 50 galaxies to look for Seyfert galaxies?

R. Weinberger: By apparent brightness, i.e. we did not observe objects brighter than about 17 mag.

Unveiling Large-Scale Structures Behind the Milky Way
ASP Conference Series, Vol. 67, 1994.
C. Balkowski and R. C. Kraan-Korteweg (eds.)

An Optical Galaxy Search in the Hydra/Antlia and the Great Attractor Region

R. C. Kraan-Korteweg and P. A. Woudt

Kapteyn Astronomical Institute, Postbus 800, 9700 AV Groningen, NL

Abstract. This is a status report of an optical galaxy search in the southern Milky Way. Our search area ($266° \lesssim \ell \lesssim 330°, |b| \lesssim \pm 10°$) lies in the general direction of the Microwave Background dipole and covers the crossing of the Galactic Plane with the Supergalactic Plane. Next to the mapping of unsuspected extragalactic structures, we want to (a) trace the proposed Hydra/Antlia Supercluster across the GP and (b) to test whether we can find any evidence for the existence of the predicted large mass concentration in the Milky Way in the Great Attractor region.

So far, we have detected over 8000 previously unknown galaxies with $D \gtrsim 0\!.\!^{'}2$. The distribution of these galaxies reveals several filamentary structures, which in general follow the adjacent unobscured galaxy distribution. In addition, two prominent overdensities are discovered: a clumpy round overdensity in Vela ($\ell \approx 280°, b \approx +6°$) and a centrally condensed overdensity close to the predicted position of the Great Attractor ($\ell \approx 325°, b \approx -7°$). The latter is centered on the cluster Abell 3627 which - with $v \approx 4300$ km/s - is at a redshift similar to that of the Great Attractor. The richness of this cluster has remained unnoticed due to the diminishing effects of the foreground extinction.

1. Introduction

A large part of the extragalactic sky is obscured by the Milky Way. This severely constrains studies of the large-scale structures in the Universe, the origin of the peculiar velocity of the Local Group (LG) and other streaming motions.

The dipole in the Cosmic Microwave Background Radiation (CMB) is explained by the peculiar motion of the LG due the gravity field of its surrounding irregular mass distribution. The whole-sky extragalactic light distribution (cf. Fig. 2 and Fig. 3 in Lynden-Bell and Lahav, 1988) emphasises the large fraction of the sky which is blocked to our view by the foreground obscuration of our Milky Way ($\approx 25\%$). It also is suggestive of various continuations of extragalactic large-scale structures across this Zone of Avoidance (ZOA). Kolatt et al. (1994) have shown that the gravitational acceleration of the LG is strongly affected by the mass distribution in the ZOA. The dipole direction changes by $31°$ if the mass distribution in the ZOA is not accounted for. It is therefore important to map these structures to their full extent, hence also in the Milky Way.

The southern Milky Way is especially exciting:

- We see traces of the nearby Puppis filament, the proposed Hydra/Antlia Supercluster and the crossing of the Supergalactic Plane with the GP – named the dinosaur's foot by Lynden-Bell (1994).

- The dipole anisotropy is located close to the southern Milky Way ($\ell = 280°$, $b = 27°$). Subtracting the decelerative component due to the Virgo Cluster, it lies even closer to the GP ($\ell = 274°$, $b = 12°$, Sandage et al., 1984, Shaya, 1984) and points to the suggested Hydra/Antlia Supercluster.

- The infall pattern of the observed peculiar velocities of elliptical galaxies was interpreted by Dressler et al. (1987) as being due to the existence of a hypothetical Great Attractor. At that time Lynden-Bell et al. (1988) deduced a mass of $\sim 5 \times 10^{16} M_\odot$ and positioned it at $l = 307°$, $b = 9°$ and $v \sim 4400$ km/s in redshift space. More recent calculations, based on a larger data set (elliptical *and* spiral galaxies) and the potential reconstruction method POTENT (Kolatt et al., 1994) now finds the center of the potential field at the same distance but slightly below the GP ($\ell \approx 320°$, $b \approx 0°$).

- It is likely that yet undiscovered members of the nearby Centaurus A group ($v = 273$ km/s) are located behind the ZOA. As shown by Kraan-Korteweg (1992), nearby 'normal' galaxies can contribute significantly to the peculiar motion of the LG. The detection of further group members would be relevant for the dynamics of the LG as well.

For these reasons, we have started a deep optical galaxy search in the southern Milky Way. We started in the extension of the Hydra/Antlia Supercluster, i.e. the general direction of the CMB dipole (Kraan-Korteweg, 1990,1991,1992). We are currently extending this search towards the Great Attractor region. If the predicted mass concentration exists and is in the form of galaxies, we should be able to retrieve a significant part of it.

In this report, we present the latest results from this galaxy search. We will give a description of our search method and discuss the properties of the detected galaxies. The unveiled galaxy distribution will be examined in context to the foreground obscuration and in combination with known extragalactic large-scale structures. Their distribution in 3-dimensional space is discussed in the following contribution (Kraan-Korteweg et al., 1994, these proceedings).

2. The Search Method

The tools for this galaxy search are very simple. It comprises a viewer with the ability to magnify 50 times and the IIIaJ film copies of the ESO/SERC survey. The viewer projects an area of 3.5×4.0 on a screen, making the visual, systematic scanning of these plates quite straightforward and comfortable.

Even though galactic extinction effects are stronger in the blue, the IIIaJ films are chosen over their red counterparts. A careful inspection between the various surveys demonstrated that the hypersensitized and fine grained emulsion of the IIIaJ films go deeper and show more resolution. Even in the deepest extinction layers of the ZOA, the red films were found to have no advantage over the IIIaJ films.

We imposed a diameter limit of $D \gtrsim 0\overset{\prime}{.}2$. Below this diameter the reflection crosses of the stars disappear, making it hard to to differentiate consistently between stars or blended stars and faint galaxies.

The positions of all the galaxies are measured with the Optronics at ESO in Garching. The accuracy of these positions is about $1''$.

For every galaxy we record the major and minor diameter, an estimate of the average surface brightness and the morphological type of the galaxy. From the diameters and the average surface brightness a magnitude estimate is derived. The reliability of the recorded diameters and the apparent magnitude are discussed in section 3.

The search was performed in two parts:

1. The Hydra/Antlia Region: It covers the area $266° \lesssim \ell \lesssim 296°$ and $-10° \lesssim b \lesssim +8°$. This region of approximately $400\square°$ encompasses 18 fields of the ESO/SERC survey (F91-F93, F125-F129, F165-F170, F211-F214).

   In the Hydra/Antlia region 2823 certain and 456 possible galaxies were discovered. Among these 3279 identifications only 97 galaxies were previously recorded by Lauberts (1982).

2. The Great Attractor Region: This is an ongoing extension towards the Great Attractor. It will extend out to $\ell \leq 330°$. So far, 24 fields have been examined (F62-F67, F94-F99, F130-F136, F171-F175) covering an area of approximately $500\square°$ within $296° \lesssim \ell \lesssim 326°$ and $-10° \lesssim b \lesssim +10°$.

   In this area 4856 galaxies are found. The percentage of previously recorded galaxies by Lauberts (1982) is similiar to the Hydra/Antlia region. Lauberts identified 122 galaxies with $D \gtrsim 1\overset{\prime}{.}0$ in this region.

3. Characteristics of the Detected Galaxies

The reliability of the magnitude and diameter estimates has been assessed by Kraan-Korteweg (1990). For one field (F213) MacGillivray extracted the images of the galaxies identified by us with COSMOS. For the larger galaxies some overlap with the ESO-LV catalog (Lauberts and Valentijn, 1989) allowed a comparison. A surprisingly good relation is found for the estimated magnitudes, with no deviations from linearity even for the faintest galaxies, and a scatter of only $\sigma = 0^m\!.47$.

The mixture of galaxy types in our survey – (E–S0 : S–I : unclassified) = (10% : 60% : 30%) – is consistent with most optical surveys. Due to the locally varying extinction it is difficult to give a homogeneous galaxy classification (the differentiation between, for instance, the bulge of a spiral galaxy and an early type galaxy is difficult in very obscured regions).

The distribution of the in this region observed diameters and the observed magnitudes is displayed in figure 1. The detected galaxies are on average quite small ($<D> = 0\overset{\prime}{.}4$) and faint ($<B_J> = 18^m\!.0$). However, extinction plays an important role in this histogram: galaxies are diminished by at least 1^m of foreground extinction at the highest latitudes ($|b| \approx 10°$). These effects increase

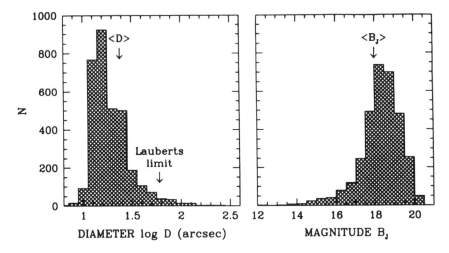

Figure 1.   The distribution of the observed diameters (left panel) and magnitudes (right panel) of the 3279 galaxies discovered in the Hydra/Antlia search.

considerably as we approach the galactic equator (cf. Fig. 2 in Kraan-Korteweg, 1990). Even so, the histogram displayed here gives a fairly good indication of the average properties and the completeness of the galaxy survey.

Accurate extinction measurements are not yet available. Using galactic HI-column densities as a tracer of the extinction and the formalism derived by Cameron (1990), rough corrections for the obscuration effects can be applied.

About 4.1% of the galaxies are listed in the IRAS Point Source Catalogue. A discussion of the cross-identification with the IRAS PSC is given in Kraan-Korteweg (1991).

4.   The Distribution of the Galaxies

The distribution of the 8135 optically detected galaxies is shown in galactic coordinates in the upper panel of figure 2. The solid line encloses the area searched to date. Next to the expected dependence of the number density on galactic latitude, strong variations with galactic longitude are evident. Are these density fluctuations due to the varying foreground extinction or are they extragalactic of origin? To answer this, we need detailed information about the obscuration.

4.1.   The Foreground Extinction

On the assumption that the gas-to-dust ratio is constant, the galactic HI will provide a rough approximation of the galactic foreground extinction. With this in mind, the HI-column densities (Kerr et al., 1986) were superimposed on the

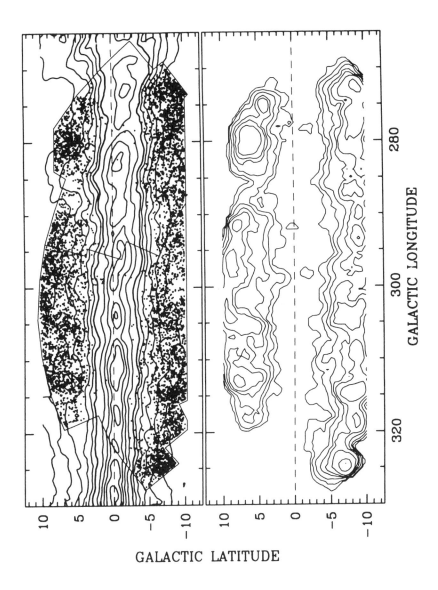

GALACTIC LATITUDE

Figure 2. Upper panel: The distribution in galactic coordinates of 8135 optically detected galaxies. Superimposed are the HI column densities from Kerr et al., 1986. The contour levels are 2.0, 2.8, 3.6, 4.5, 5.5, 6.5, 9.0, 11.5, 14.0, 16.5 and 20×10^{21} atoms/cm^2. The thick line corresponds to a HI-column density of 6.5×10^{21} atoms/cm^2. Lower panel: The isopleths as derived from the galaxy distribution. The contour levels are 0.5, 2, 5, 10, 15, 25, 50, 75 and 100 galaxies/□°.

galaxy distribution in Fig. 2. Note that the asymmetry of the HI column densities with respect to the galactic equator is reflected in the galaxy distribution obtained here (265° - 300°). Overall, HI seems to be a more reliable tracer of extinction down to very low galactic latitudes than previously assumed. Comparing, for instance, the lowest galaxy density contour (0.5 galaxies/□°, lower panel of Fig. 2) with the thick HI-contour (6.5×10^{21} atoms/cm^2, upper panel of Fig. 2) shows them to be remarkably similar. Below this column density – which corresponds roughly to an extinction of $A_B \sim 5^m$ – hardly any galaxies are visible.

At lower galactic latitude the HI column density is not a reliable tracer of the galactic dust. The CO distribution (e.g. Dame et al., 1987) can be used there as a tracer of dust. This is not relevant for our study as we have only very few galaxies that close to the dust equator.

The galactic dust clouds generally have a patchy distribution and can locally cause heavy obscuration. A comparison of the galaxy distribution with the dark cloud distribution (Feitzinger and Stüwe, 1984) indicates that the observed overdensities and filaments in the galaxy distribution can not be explained by extreme transparent regions. Therefore, the galaxy overdensities and filaments – which stand out quite markedly in both panels of Fig. 2 – are likely to be of extragalactic origin.

4.2. Structures in the Galaxy Distribution

In the following we will discuss the structures (filaments, overdensities, voids) found in the galaxy distribution. To follow the possible continuity of these structures with extragalactic features above and below the ZOA, we have blended our galaxy distribution (D $\gtrsim$ 0.2) with the distribution of Lauberts galaxies (D $\gtrsim$ 1.0, Lauberts, 1982) in figure 3.

The very round but clumpy overdensity at $\ell = 280°, b = +6°$, the Vela Cluster, has first been emphasized by Kraan-Korteweg and Woudt (1994). It is located in the extension of the Hydra (270°,+27°) and Antlia (273°,+19°) clusters. The galaxies in the overdensity around (275°,−9°) are on average quite small and are likely to be more distant than the galaxies in the Hydra/Antlia extension.

The two main filaments above the GP around $\ell \approx 305°$ and $\ell \approx 313°$ suggest the continuation of the Centaurus (302°,+22°) – Pavo (332°,−24°) Supercluster (Fairall, 1988) into the GP. The filament below the GP at $\ell \approx 314°$, however, has no counterpart in the galaxy distribution beyond the ZOA. Given the relative small diameters of the galaxies in this filament, it most likely is not connected with the Centaurus – Pavo Supercluster.

The most prominent overdensity in this galaxy search is located at $\ell = 325°$ and $b = -7°$. It is centered on the nearby ($cz \approx 4300$ km/s) Abell cluster A3627 in the constellation of Triangulum Australis. These galaxies are on average quite large (D $\sim$ 50″) with a large fraction of early type galaxies. This cluster lies on the border of our search area. Inspection of the neighbouring field – which is presently being surveyed – indicates that we have uncovered only half of the cluster. Based on the morphology and the sizes of these galaxies, and the angular extent of this cluster (about 10 degrees on the sky if the adjacent field

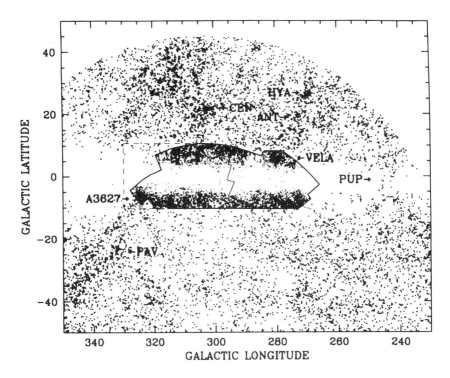

Figure 3. The distribution of the detected galaxies (small dots) in connection with known galaxies in this area (big dots) from Lauberts (1982). The solid line outlines our search area, the dashed line shows the planned extension.

is included), it is clear that we have discovered a major cluster of the nearby Universe.

Due to the diminishing effects of the foreground extinction ($A_B \approx 1^m - 2^m$) the richness of this centrally condensed nearby cluster has never been noted, even though this cluster lies within a few degrees of the predicted mass concentration, the Great Attractor.

5. Conclusions

Despite the strong obscuration effects in optical wavebands, deep optical galaxy searches have been proven to be good tracers of extragalactic structures down to quite low galactic latitude. An advantage of optical galaxy searches over, for instance, IRAS and HI studies, is the fact that we retrieve all galaxian types, thus probing nearby structures through spirals and irregulars (groups and filaments), and also the cores of dense groups and rich clusters through the early type galaxies.

We find the HI column densities to be a fairly good tracer of the total foreground extinction. The low number density isopleths match the high HI column densities rather well (cf. N = 0.5 galaxies/$\Box^\circ$ and N(HI) = 6.5×10^{21} atoms/cm^2 in Fig. 2). In collaboration with B. Burton (Leiden), we have started a new program to obtain accurate extinction values in the ZOA through 2-color photometry and the Mg$_2$-index of early-type galaxies.

Several extragalactic filaments and overdensities have been identified. A round but clumpy overdensity, the Vela Cluster ($280^\circ,+6^\circ$), is located in the Hydra/Antlia extension. Whether this is part of the Hydra/Antlia Supercluster will be established with redshift data (cf. next contribution). The filaments above the GP probably belong to the Centaurus – Pavo Supercluster. A general underdense region is seen around ($298^\circ,-8^\circ$); it is not yet clear whether this underdense region is part of a void.

Within $\sim 10^\circ$ of the predicted Great Attractor, we have uncovered a major cluster at a redshift of ~ 4300 km/s, i.e. the approximate velocity-distance of the Great Attractor. This cluster has been identified by Abell as A3627. However, the prominence of this cluster was hidden by the veil of the Milky Way.

At this point in time we can not assess whether this rich cluster with its many ellipticals is massive enough to explain the observed infall motion into this general area. However, we have started a thorough analysis of this cluster. This will entail an extension of the search area to determine its true extent on the sky, redshift measurements of the cluster members and CCD-photometry. These observations will allow a proper analysis and mass estimate of this cluster, from which we can deduce the gravitational perturbation on its surroundings. This will be compared with the observed peculiar velocity field.

Acknowledgments. We kindly thank H. Bloemen (Leiden) for providing the integrated galactic HI data as derived from the measurements by Kerr et al. (1986). PAW thanks the LKBF for their financial support, which allowed him to participate at this meeting. The research of RCKK has been made possible by a fellowship of the Royal Netherlands Academy of Arts and Sciences.

References

Abell, G. O., Corwin, H. G., & Olowin, R. P. 1989, ApJS, 70, 1

Cameron, L. M. 1990, A&A, 233, 16

Dame, T. M., Ungerechts, H., Cohen, R. S., de Geus, E. J., Grenier, I. A., May, J., Murphy, D. C., Nyman, L.-Å., & Thaddeus, P. 1987, ApJ, 322, 706

Dressler, A., Faber, S. M., Burstein, D., Davies, R. L., Lynden-Bell, D., Terlevich, R. J., & Wegner, G. 1987, ApJ, 313, L37

Fairall, A. P. 1988, MNRAS, 230, 69

Feitzinger, J. V., & Stüwe, J. A. 1984, A&AS, 58, 365

Kerr, F. J., Bowers, P. F., Jackson, P. D., & Kerr, M. 1986, A&AS, 66, 373

Kolatt, T., Dekel, A., & Lahav, O. 1994, submitted to MNRAS

Kraan-Korteweg, R. C. 1990, in Reviews in Modern Astronomy 2, G. Klare, Heidelberg: Springer, 119

Kraan-Korteweg, R. C. 1991, in Large-Scale Structures and Peculiar Motions in the Universe, D. Latham and L.N. da Costa, San Francisco: PASP, 165

Kraan-Korteweg, R. C. 1992, in The 2nd DAEC Meeting on the Distribution of Matter in the Universe, eds. G. Mamon and D. Gerbal (Meudon), 202

Kraan-Korteweg, R. C., Cayatte, V., Balkowski, C., Fairall, A. P., & Henning, A. P. 1994, these proceedings

Kraan-Korteweg, R.C., & Woudt, P.A. 1994, in 9th IAP Astrophysics Meeting and 3rd Meeting of the EARA on Cosmic Velocity Fields, eds. F. Bouchet and M. Lachièze-Rey, 557

Lauberts, A., & Valentijn, E. 1989, The Surface Photometry Catalogue of the ESO/Uppsala Galaxies, Garching: ESO

Lauberts, A. 1982, The ESO/Uppsala Survey of the ESO(B) Atlas (Garching: ESO)

Lynden-Bell, D. 1994, these proceedings

Lynden-Bell, D., Faber, S., Burstein, D., Davies, R. L., Dressler, A., Terlevich, R., & Wegner, G. 1988, ApJ, 326, 19

Lynden-Bell, D., & Lahav, O. 1988, in Large Scale Motions in the Universe, V.C. Rubin and G.V. Coyne, Princeton: Princeton University Press, 199

Sandage, A., & Tammann, G. A. 1984, in Large Scale Structures of the Universe, Cosmology and Fundamental Physics, First ESO-CERN Symp., G. Setti and L. van Hove, Garching: ESO, 127

Shaya, E. J. 1984, ApJ, 280, 470

Discussion

E. Meurs: You gave the numbers of galaxies found by Lauberts (in his ESO catalogue) and by yourself (many more). You also mentioned the much larger diameter cut-off he had adopted. It would then be interesting to know how complete Lauberts is to his diameter cut-off of 60 arcsec.

P. Woudt: Of the 8135 galaxies detected in our optical galaxy search 219 are Lauberts galaxies. In the mean this corresponds to the number of galaxies we find with $D \geq 60$", i.e. the diameter limit of the Lauberts catalogue. However, 20% of the Lauberts galaxies are actually smaller than 60" according to our diameter measurements. On the other hand, we find that 25% of the galaxies larger than 60" were not previously recorded by Lauberts.

The 3-Dimensional Galaxy Distribution in the ZOA from Hydra/Antlia to the Great Attractor Region

R.C. Kraan-Korteweg

Kapteyn Astronomical Institute, Postbus 800, 9700 AV Groningen, NL

V. Cayatte and C. Balkowski

Observatoire de Paris-Meudon, D.A.E.C., 92195 Meudon, France

A.P. Fairall

Dept. of Astronomy, University of Cape Town, Rondebosch, 7700 SA

P.A. Henning

*University of New Mexico, Dept. of Physics and Astronomy,
Albuquerque, NM 87131*

Abstract. Three observational programs are ongoing to determine the
3-dimensional distribution of the galaxies detected in our galaxy search:
multifiber-spectroscopy in the densest regions, optical spectroscopy of
more isolated galaxies, and HI-observations of fairly large, low surface
brightness galaxies. The three approaches are complementary in the
depth of the volume they cover and the galaxy population they are optimal for. The merits and results of the different methods are discussed.

So far over 600 new redshifts have been reduced. We have good cov-
erage within the area $270° \leqslant \ell \leqslant 306°, |b| \leqslant 10°$. The structures unveiled
here are discussed. The filamentary extension, proposed earlier, from the
Hydra cluster $(270°, +28°, \sim 3300$ km/s) is confirmed and can be traced
across the ZOA to $(280°, -10°, \sim 2500$ km/s). The Vela overdensity at
$(280°, +6°)$ is a chance superposition on the Hydra/Antlia filament. It is
part of an overdensity (supercluster?) at $v \sim 6000$ km/s which might be
linked with the Great Attractor through a very broad filament.

Very distant sheets are disclosed as well. One is suggestive of a
connection between the Horologium clusters and the Shapley clusters.
Given its approximate distance (~ 16000 km/s) and its extent on the sky
(roughly 100°), this would imply an extremely large structure.

1. Introduction

As described in the previous chapter (Kraan-Korteweg & Woudt 1994, [KKW]),
various filaments and overdensities have been uncovered in the southern Milky
Way. To assess their true dimensions and their gravitational effects on the
surroundings, we need to know their distribution in space. As a first step in

this direction, we have begun to measure redshifts of a representable part of the uncovered galaxies. In the following, these observing programs are described and the results obtained so far discussed. It should be stressed that this is an ongoing project. Our search started at $\ell \sim 266°$ and we recently progressed to the Great Attractor (GA) region. This is evident also in our resulting redshift data. We have good redshift coverage within $270° \lesssim \ell \lesssim 306°$, $|b| \lesssim 10°$, whereas we are presently extending our redshift measurements to $\ell \lesssim 330°$ (cf. Fig. 1).

2. The Observational Redshift Programs

2.1. Multifiber Spectroscopy in Dense Regions (ESO)

As illustrated by KKW in the lower panel of Fig. 2, the number density in the galaxy distribution varies strongly with location. The densest structures have densities above 25-50 galaxies/$\square°$, i.e. optimal for multifiber spectroscopy. We have used the multifiber systems mounted on the 3.6m telescope of ESO, i.e. Optopus and the follow-up instrument Mefos with scientific test time. Further details on Mefos and first results are described by Felenbok 1994, and Cayatte, Balkowski & Kraan-Korteweg 1994, both in this volume. The main characteristics of these 2 instruments are:

Optopus   Up to 52 objects can be observed simultaneously within a circular field of $D \leq 33'$. Usually 4-5 of the object fibers were positioned on the sky. The fibers are plugged manually into prepunched plates at the telescope. As the plates have to be prepared well in advance with a predetermined observing schedule, this allows for little flexibility during the run, or re-use of the plates during a following observing run.

In 2 runs (1990, 1992) we have observed 500 galaxies on 23 fields with Optopus. 19 fields were reduced at the time of this workshop. Of the 407 observed galaxies we have obtained 271 (66%) reliable and 44 (11%) possible redshifts. For 79 (19%) of the objects the S/N was too low to extract a redshift. 13 (3%) of the spectra were polluted by the light of a superimposed star.

Mefos   Mefos is mounted in the prime focus and 29 objects can be observed simultaneously within a field of $D \leq 60'$. The fibers are located in moveable arms in a circle around the field. The positioning of the arms is initiated at the telescope by a computer program. Each arm contains an image fiber and 2 spectral fibers. A short exposure with the image fiber allows exact positioning of the spectral fiber. The second spectral fiber is offset by 60″ giving one sky exposure per object.

With Mefos we have observed 162 galaxies on 9 fields (1993). 6 fields have been reduced: out of the 113 observed galaxies, 92 (82%) have reliable, and 7 (6%) have possible redshifts. 8 (7%) spectra have too low S/N, whereas 6 (5%) were useless due to foreground stars.

In principle, the overhead and the efficiency of the two instruments are comparable. The limiting magnitude for both instruments for a 1^h-exposure is $B_J \approx 18^m_.5-19^m_.0$. However, it seems irrefutable that the Mefos observations with its accurate, on-line pointing and improved sky subtraction result in a rel-

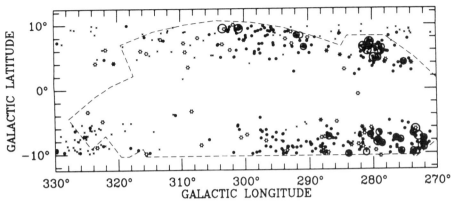

Figure 1.   The distribution of the galaxies with redshifts. Open circles mark the multifiber fields (small: Optopus, large: Mefos) in which
galaxies with reduced spectra are entered as small dots; filled dots: redshifts obtained at the SAAO; open stars: redshifts obtained at Parkes;
small crosses redshifts measured by others.

atively larger fraction of accurate redshift determinations (82% versus 66%).
The distribution of the observed fields and the reduced redshifts are displayed
in Fig. 1.

2.2.   Individual Spectroscopy of 'HSB' Galaxies (SAAO)

With the above observations, only the highest density peaks are localised in
space (compare e.g. the multifiber fields in Fig. 1 with the galaxy distribution
in Fig. 2 of KKW). A more homogenous redshift coverage over the whole search
area is required to fit these parts of the puzzle into a larger framework. We therefore started a complementary program (Kraan-Korteweg, Fairall, Balkowski &
Woudt) to individually obtain optical spectra of all the brighter galaxies. For
this we used the 1.9m telescope of the SAAO. On average we obtained good
S/N spectra within a reasonable exposure time (typically $10 - 30^{min}$ to at most
1^h) for galaxies down to a limiting magnitde of $B_J \approx 17^m.0-17^m.5$ – if they have
high 'central' surface brightness. A detailed description of these observations is
presented in Kraan-Korteweg, Fairall & Balkowski, 1994.

In the 5 weeks of observing time allocated to us between 1991 and 1993, we
observed 292 galaxies (mainly $\ell \lesssim 306°$). For 237 (81%) we have good redshifts
plus an additional 18 (6%) with possible redshifts. Here again 5 (2%) spectra
had to be discounted due to foreground stars and 32 (11%) due to too low S/N.
The galaxies with good redshifts are entered into Fig. 1 as filled dots.

2.3.   21cm Observations of LSB Extended Spiral Galaxies (Parkes)

Although we aimed at complete coverage to a certain magnitude and diameter
limit, we did not succeed with the optical spectroscopy: a significant fraction of
apparent bright galaxies cannot be traced in this manner. In general, this concerns nearby spirals (and also dwarfs) which, seen through a layer of extinction,

are extended, very low surface brightness objects. However, these objects are important, as they are the dominant constituents of the nearby (filamentary) galaxy population.

In order to determine their redshifts, we started observing these LSB galaxies with the 64m radio telescope in Parkes in 1993 (Kraan-Korteweg, Henning, Schröder and van Woerden). In this period, we observed 159 galaxies without previous redshift estimates. We worked in total power mode, with the 1024 channel autocorrelator and a bandwidth of 32 MHz centered at 3000 km/s, resulting in a good sensitivity for the velocity range of 600 $<$v$<$ 5400 km/s. Some of the non-detections were reobserved with a higher central velocity. General integration times were 30^{min} on the source. A number of stronger sources were detected in 10^{min} integration time.

Our success rate was quite high: we obtained 65 reliable, 5 marginal and 12 possible detections. Considering that a large part of our daytime observations did preclude a detection due to high-amplitude standing waves in the spectrum, this results in a *detection rate of over 50%*. The detected galaxies are entered as starred symbols into Fig.1. Note that the HI-detections finds galaxies deeper in the obscuration layer compared to the optical observations.

2.4. Typical Properties of the Observed Galaxies

In total, we derived 670 redshifts for 632 galaxies (38 galaxies were observed in more than one program) within $270° \leqslant \ell \leqslant 330°, |b| \leqslant 10°$, with good coverage for $\ell \leqslant 306°$. In addition, redshifts for 137 galaxies within this area had been determined previously. These are extracted from the Southern Redshift Catalogue (Fairall & Jones 1991, [SRC]), and the IRAS 2Jy Redshift Survey (Strauss et al. 1992, [IRAS RS]) and plotted as small crosses in Fig. 1.

Diameter and Magnitude Distributions   The applied observing methods differ strongly and before we start to analyse the redshift distribution, it is important to understand more about the properties of the observed galaxies. In figure 2 we have plotted the distribution of the diameters and magnitudes (uncorrected for foreground extinction) separately for the different observing methods. The total sample is displayed in the bottom panel. The mean of the individual samples is indicated in the respective panel. It should be kept in mind, that we are looking at galaxies which are diminished by $A_B \approx 1^m - 5^m$, and the intrinisic diameters/magnitudes therefore are considerably larger/brighter.

The distribution for the Optopus and Mefos data are quite similar. Only a small number of large, bright galaxies are observed per field – a reflection of the low density of bright galaxies per multifiber field. The distribution gradually increases to a peak at $12'' - 20''$ and $17^m.5-19^m.0$; it is representative of the typical properties of the overall galaxy search (cf. Fig. 1. in KKW).

With the SAAO observations we are probing the bright end of the luminsity function. It should furthermore be taken into account that the brightest galaxies in the ZOA generally have already been observed by others. Even so, the mean 'obscured' diameter and magnitude is quite large ($52''$) and bright ($16^m.1$).

The diameter and magnitude distribution of the HI detected galaxies – with a mean of the parameters of $<$D$>=$ $68''$ and $<$B$_J$ $>=15^m.6$ – clearly demon-

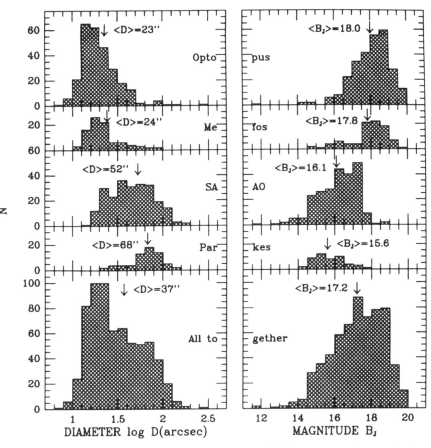

Figure 2.   The distribution of the observed diameters and magnitudes of galaxies, with redshifts measured by us, divided according to the applied observing method.

strate the importance of the HI-observations as complementing the optical observations.

It can be maintained that the three observing methods supplement each other quite well: the SAAO and HI-observations together cover the bright end of the galaxies detected in the ZOA. In addition, the multifiber spectroscopy gives a good description of the clusters in the ZOA as well as a good representation of the overall galaxy sample. In the Hydra/Antlia search area – which has good velocity coverage – we are, for instance, 96% complete for $B_J < 15^m.0$, 82% for $B_J < 16^m.0$ and nearly 70% for $B_J < 17^m.0$. We are still lacking redshift information for some extended LSB objects. This is reflected in the lower completeness percentages as a function of observed diameter: to date we are 'only' 81% complete for galaxies with $D \geq 60''$ (70% for $D \geq 45''$, but still 40% for $D \geq 30''$). These could be fairly local dwarfs or even misidentified galactic objects.

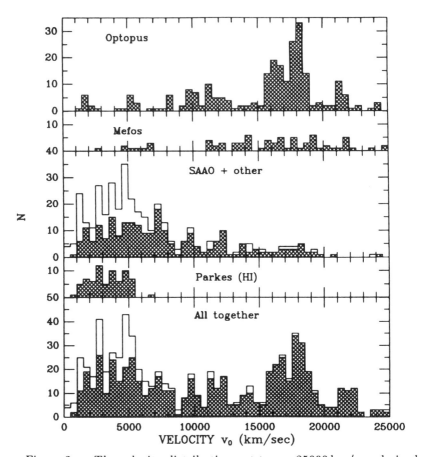

Figure 3.   The velocity distribution out to $v_0 < 25000 \, \mathrm{km/s}$ as derived
with the different observing methods. Redshifts published in the literature are added to the SAAO-panel. For the Optopus data, 18 further
galaxies (7%) have $v > 25000 \, \mathrm{km/s}$ and 17 (18%) for Mefos.

Redshift Distributions   The redshift distributions are illustrated in an equivalent way in Fig. 3. Redshifts from the literature are added to the SAAO panel,
as these are galaxies which would have entered this observing program had they
not been observed before.

The differences in the velocity histograms are quite conspicuous. for which
the origin clearly is found in the characteristics of the chosen observing methods.

The velocity distributions of the multifiber systems are quite similar. In
the selected fields, only a few galaxies lie nearby, while there is a large fraction
of distant galaxies ($v \gtrsim 10000 \, \mathrm{km/s}$). In addition, 18 (7%) of the Optopus and
17 (18%) of the Mefos galaxies (not shown here) are even more distant (they
seem to cluster around 34000 km/s). The distinct peaks in this distribution
must be due to the fact that we did pick up clusters in these overdense areas. It
should be noted that these same peaks are repeated in the SAAO-observations,

although less pronounced. The Mefos distribution covers the same velocity range as Optopus but is considerably flatter. Whether this has its origin in different extragalactic structures within the selected fields, or the fact that the Mefos fields cover a larger area and therewith pick up the more homogenously distributed background galaxies (remember that the fraction of high velocity galaxies is larger compared to Optopus) cannot be answered yet due to the small number of reduced Mefos fields to date.

In correspondence to the properties of the SAAO galaxies, these galaxies are good tracers of the 'nearby' Universe (v≲10000 km/s), with some indication of structures out to 20000 km/s. The Parkes detections again emphasize that we can retrieve a large part of the very nearby (v≲5000 km/s) galaxy population only with the HI-observations.

The combined velocity distribution is displayed in the bottom panel. As discussed in this section, the three observing methods are complementary and together provide a representative sample for the galaxy distribution in the ZOA.

3. The Resulting Distribution in Redshift Space

We will now look at the distribution in space to outline the extragalactic structures responsible for the velocity peaks, to see how these structures fill the gap in the ZOA, and how they connect to the structures adjacent to the ZOA. The SRC (Fairall & Jones, 1991) and the 2Jy IRAS-RS (Strauss et al., 1992) are used to outline the structures next to the Milky Way. Caution is demanded when interpreting the combined distributions. Not only do we have the problems of the foreground extinction and the selection effect due to the chosen observing techniques, but the SRC constitutes an uncontrolled data set: it lists redshifts in the southern hemisphere 'purely' on the basis of their availability.

3.1. Nearby Structures

Nearby structures are clearly recognizable in skyplots with discrete velocity intervals. The redshift data in our search area and its surroundings are plotted in such a way in figure 4: for 500<v<3500 km/s in the upper panel, and 3500<v<6500 km/s in the lower panel. Within these panels the velocities can be identified to $\Delta v=1000$ km/s through the symbols. For the orientation the reader is referred to Fig. 3 in KKW in which the main features are marked.

Note the prominence of the dinosaur's paw in the upper panel. The new data adds important points in confirming and outlining the suspected Hydra/Antlia-extension, i.e the middle feature of the 'paw' ($280°>\ell>270°$). It seems to be a filamentary structure interspersed with groups and spiral-rich clusters reaching from Hydra ($l = 270°, b = 28°$, v∼3300 km/s) to Antlia ($273°, 19°$, ∼2800 km/s) entering the most opaque part of the Milky Way at $b \sim 3°$, emerging on the opposite side with a previously unknown group at ∼280°, ∼-7°,∼2500 km/s. Whether it stops at the border of our search area or continues even further is uncertain.

Next to that, we see a distinct filament which rises from Puppis ($\ell = 245°, b = 0°$, v∼1500 km/s) towards Antlia (redshifts in this part of the ZOA result from an investigation by Kraan-Korteweg & Huchtmeier, 1992). This filament continues across Antlia towards Centaurus where it intersects (or con-

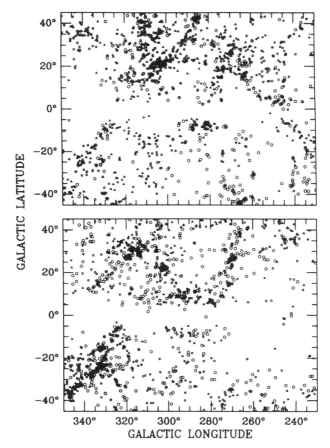

Figure 4.   Extragalactic 'nearby' (500 < v_0 < 6500 km/s) large-
scale structures across the Milky Way. Different velocity intervals are
marked by different symbols (top: ○ 500-1500, □ 1500-2500, △ 2500-
3500 km/s, bottom: ○ 3500-4500, □ 4500-5500, △ 5500-6500 km/s).

tinues?) with the third filament of the dinosaurus' paw, which then folds back
crossing the Milky Way once again. Note furthermore the overabundance of
very nearby galaxies (circles) in the lower, right-hand quarter of this skyplot.

Overall the impression is of a cellular structure, reminiscent of the smaller
structures within larger walls as visualized in the video shown at the workshop
by Fairall at al. (1994) and discussed in this volume.

This is evident also in the lower panel. Due to the velocity dispersion in
groups and clusters, some of the clusters show up again, as well as the slightly
more distant, narrow filament which leads from the Centaurus cluster back across
the GP, over our prominent overdensity (A3627) to about 340°, −25°. Apart
from that we see a striking overabundance of galaxies with v∼ 4000 km/s (open
circles) within broader structures. The most interesting is the broad filament
which starts below the GP at about 330°, −30°, ∼4500 km/s ('adjacent' to the

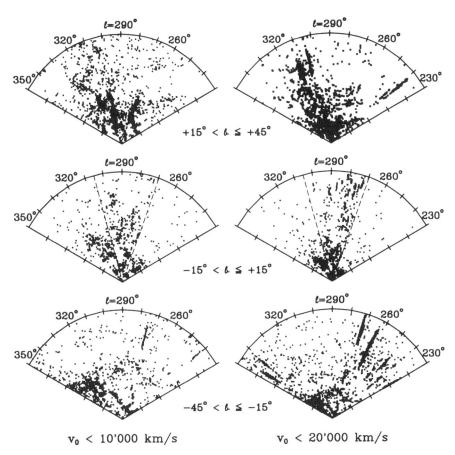

$v_0 < 10'000$ km/s $v_0 < 20'000$ km/s

Figure 5.   Redshift slices out to $v_0 < 10000$ km/s (left panel) and $v_0 < 20000$ km/s (right panel) for galactic longitudes $230° < \ell < 350°$) for 3 adjacent latitude slices of thickness $\Delta b = 30°$ (top: above the GP; center: in the GP; bottom: below the GP.

narrow filament at ~ 3500 km/s that connects to Centaurus), crossing the GP and surfacing again at slightly higher velocities around $\ell \sim 295°$. From there it seems to merge with the Vela-overdensity which peaks at v~ 6000 km/s. The Vela-overdensity is therefore not – as anticipated earlier – a main part of the Hydra/Antlia filament, but a coincidental superposition of 2 structures.

The broad feature is even more impressive in the left panel of Fig. 5 in which redshift cones out to 10000 km/s are displayed: from below the plane (bottom panel, left) we can trace the structure at 330°, 4000 km/s into the Milky Way (middle panel, left) where it merges with a shallow but large-sized overdensity centered on $\sim 290°$, ~ 6000 km/s. This includes the above mentioned Vela-overdensity as well as another recently investigated cluster within this area (not shown here) at $\ell = 280°, b = 11°$, v~ 6000 km/s (Stein, 1994). It is interesting that Saunders et al. (1991) already predicted a supercluster (S4) at 6000 km/s in this part of ZOA, although it was only weakly significant.

Is this shallow overdensity part of the predicted GA as suggested by the data here? With a newly developed reconstruction algorithm of the perturbation field using IRAS galaxies, Hoffman (these proceedings) predicts three peaks in the soutern Milky Way, the Hydra/Antlia-Centaurus Supercluster at $\ell \sim 300°$, ~ 3000 km/s, the GA at $\ell \sim 325°, \sim 4500$ km/s, and the Vela-overdensity at $\ell \sim 280°, \sim 6000$ km/s, all of which are embedded in a larger-scale, but lower-level overdensity. This seems to agree remarkably well with the features uncovered here.

3.2. Distant Structures

Redshift cones out to 20000 km/s are plotted in the right panel of Fig. 5. It is interesting that the peaks at 9-11000 km/s and 16-18000 km/s (cf. Fig. 3, bottom panel), which are apparent as clumpy distributions in the middle panel, are also repeated in the adjacent panels. At 16-18000 km/s we can actually link the Horologium clusters from below the plane, diagonally across the GP with the new clusters detected here, to the Shapley clusters above the plane. Likewise, though less pronounced, a connection at 9-11000 km/s is suggested in the galaxy distribution. Whether this signifies a curious line-up of structures at preferential redshifts, or is indicative of coherent wall-like structures over 100° in the sky is not clear yet (at a distance of about 16000 km/s the latter leads to sizes of nearly 300 Mpc h^{-1}!). The forthcoming data will help to unravel these questions.

4.   Concluding Remarks and Future Plans

The observing methods used so far are quite powerful in retrieving the 3-dimensional galaxy distribution in the ZOA, and the various, unveiled structures are discussed in detail in the previous section.

Since the workshop, we have had 3 further succesful observing runs. Based on these observations we expect about 700 new redshifts (~ 400 based on the Mefos observations, ~ 200 redshifts from the SAAO and ~ 90 detections of nearby spiral galaxies with the Parkes radio telescope) foremost in the extension of our search area from $306° < \ell < 330°$, resulting in an equivalent coverage as for the Hydra/Antlia region. It will take some time to reduce this large data set and to do a proper analysis. From the raw data it is clear, however, that the prominent cluster A3627 and the galaxies around it are indeed at ~ 4500 km/s. It seems more aligned with the narrow filament that merges with the Centaurus cluster, than with the broader filament – which could also be substantiated with the new data – that leads to the Vela-overdensity.

The ultimate goal is, of course, to blend our data with whole sky surveys in order to calculate the effects of the galaxy distribution in the ZOA on the peculiar motion of Local Group and explain the observed streaming motions. This requires accurate diameters, magnitudes and morphological types, and foremost good extinction determinations. With the extinction values, we can apply Cameron's law (Cameron, 1990) to correct the observed parameters for the foreground absorption, which will then lead to an assessment of the completeness as a function of positon in the ZOA.

For this reason we have started a new program doing B and R CCD-photometry with the Dutch telescope at LaSilla. High priority objects are elliptical galaxies for which we already obtained spectra with Optopus and Mefos. These spectra allow the measurement of the Mg_2-index (this can be easily achieved even for galaxies as distant as 18000 km/s). Using the relation between color and the Mg_2-index (Burstein et al. 1988), we can predict reddening values with an accuracy of $<0^m.03$.

Based on these reddenings, we can correct the observed parameters within the Mefos or Optopus field for the obscuration effects. Through interpolation - including HI-column densities - we can correct the properties of adjacent galaxies as well. This will prove extremely important for our analysis of the 2D- and 3D-distribution of the discovered galaxies in the ZOA and their connections to structures beyond the Milky Way.

Acknowledgments. The research of RCKK has been made possible by a fellowship of the Royal Netherlands Academy of Arts and Sciences.

References

Burstein, D., Davies, R.L., Dressler, A., Faber, S.M., Lynden-Bell, D., Terlevich, R., Wegner, G. 1988, in Towards Understanding Galaxies at Large Redshifts, eds. R.G. Kron & A. Renzini, p.17

Cameron, L.M. 1990, A&A, 233, 16

Cayatte, V., Balkowksi, C., & Kraan-Korteweg, R.C. 1994, these proceedings

Fairall, A.P., & Jones, A. 1991, Southern Redshifts: Catalogue & Plots, Publ. Univ. of Capetown, No. 11

Fairall, A.P., Paverd, W.R., & Ashley, R.P. 1994, these proceedings

Felenbok, P. 1994, these proceedings

Hoffman, Y. 1994, these proceedings

Kraan-Korteweg, R.C., Fairall, A.P., & Balkowski, C., 1994, submitted to A&A

Kraan-Korteweg, R.C., & Huchtmeier, W.K. 1992, A&A, 266, 150

Kraan-Korteweg, R.C., & Woudt. P.A. 1994, these proceedings

Saunders, W., Frenk, C., Rowan-Robinson, M., Efstathiou, G., Lawrence, A., Kaiser, N., Ellis, R., Crawford, J., Xia., X., & Parry, I. 1991, Nature, 349, 32

Stein. P., 1994, Ph.D. thesis, Univ. of Basle, in preparation

Strauss, M., Huchra, J.P., Davis, M., Yahil, A., Fisher, K., & Tonry, J., 1992 ApJS, 83, 29

Discussion

W. Saunders: Cameron's diameter and magnitude corrections are quite a crude piece of work. Are there any plans to do a more thorough analysis?

R. Kraan-Korteweg: As far as I am aware there are no such plans. But I fully agree that a more detailed study of the extinction corrections as a function of morphological type, bulge-to-disk ratio, inclination would be very valuable for our studies of the ZOA, especially with regard to completeness. Not only will the formalism allow corrections of the obscured parameters, it will furthermore allow us to assess to which latitude, or HI-column-density as a measure of extinction, we can expect to find galaxies down to a certain magnitude or diameter limit.

H. van Woerden: Concerning your Parkes survey I have two questions. (1) You find many large and bright galaxies with redshifts below 5000 km/s. Could you not get optical redshifts for these? (2) At Parkes about half of your galaxies remained undetected. Is it likely that their velocities fell outside your observing band of 0-6000 km/s?

R. Kraan-Korteweg: (1) With our optical spectroscopy, foremost at the SAAO, we tried to obtain spectra for all the largest and brightest galaxies. However, many of these extended spiral galaxies have a very low surface brightness and we did not succeed in obtaining a good S/N spectrum. In this respect the HI observations are a really important complementation in mapping the nearby galaxy population, as such galaxies are easily detectable with radio telescopes. (2) Our observing strategy at Parkes was as follows: in a first step we centered the 32 MHz band at $v_c=3000$ km/s, integrating up to 30 min at most. This gave us a good response for the velocity range $800 \leq v \leq 5400$ km/s. In a second step, we chose a higher central velocity, i.e. $v_c=7000$ km/s where the overlap around 5000 ensures detections of broad linewidth galaxies at ≈ 5000 km/s. Our detection rate at the lower velocity range was so high that we hardly ever used the second step. Our selection criteria for identifying spiral galaxies are quite reliable and I am therefore convinced that many of our non-detections can indeed be found at higher redshifts. In forthcoming observations at Parkes we plan to reobserve some of the non-detections at higher velocities.

N. Lu: It is a bit uncertain to claim that the two structures separated by $8° \sim 10°$ are physically related because it might well be the case that the two concentrations are separated by a void. Thus, it might be a good idea to measure redshifts of a small sample of IRAS sources in between.

R. Kraan-Korteweg: I think the new data is quite convincing. Before our galaxy search we only had the Hydra and Antlia clusters at $b=25°$ and $b=18°$ and some filamentary structure below the ZOA pointing towards Fornax. The

latter is seen foremost in 2-dimensional distributions but has been shown by Mi-
tra in 1989 to be real in velocity space as well. Our new velocity data now fills in
this filament to +5°, picking it up again at -5°. It seems a filament interspersed
with groups and spiral rich clusters, matching the structures above and below
the ZOA quite well in velocity space. I would therefore think it much more
improbable, to expect a void at the most opaque part of the ZOA. Independent confirmation that this structure continues across the ZOA has been given
by Ofer Lahav's Wiener reconstruction method of incomplete sky surveys. It
would of course be interesting to measure galaxies in the opaque part. However,
IRAS identifications in the highest obscuration layer are very uncertain due to
the confusion with galactic objects. In this case, I think it might be just as
fruitful to trace this filament across the Milky Way with a blind HI search.

C. Balkowski: Did you look for a periodicity of about 120 Mpc in the velocity distribution like Broadhurst found in a pencil beam or Vettolani *et al.* in
a redshift survey in the galactic pole region?

R. Kraan-Korteweg: As most of our galaxies have redshifts below 25,000
km/s (only 3-4% have V>25,000 km/s) our galaxy sample is not deep enough to
find a periodicity of that order. A typical velocity spacing between peaks below
25,000 km/s is of the order of 5,000 km/s, but quite irregular and not distinct.
This seems to correspond to the typical sizes of voids.

MAPPING LARGE SCALE STRUCTURE BEHIND THE GALACTIC PLANE

RON MARZKE and JOHN HUCHRA

Harvard-Smithsonian Center for Astrophysics, 60 Garden Street, Cambridge, MA 02138, USA

ABSTRACT   We have extended the CfA survey to low galactic latitudes by observations of optically selected galaxies in the Zwicky and Nilson catalogs. We have done this to investigate the relation between the Great Wall in the north Galactic cap to the Perseus-Pisces chain in the south Galactic cap. These structures are *not* simply connected across the Zone of Avoidance. These structures, which appear to be coherent on scales up to ~ 100 h^{-1} Mpc, occur on the boundaries of neighboring voids of considerably smaller scale < 50 h^{-1} Mpc.

INTRODUCTION

In the optical, our own Galactic plane significantly obscures nearly half the sky. Optical redshift surveys are typically restricted to galactic latitudes $b \geq 30$, where the extinction in the blue is usually less than a few tenths of a magnitude. As luck would have it, many of interesting features of the galaxy distribution lie very close to the Galactic plane.

In both Galactic caps, the CfA Redshift Survey has revealed large, coherent structures which span the surveyed volume (Geller and Huchra 1989). The sample in the north galactic cap includes the Great Wall, the southern sample the Perseus-Pisces chain. Note that the Second Southern Sky Redshift Survey (SSRS2, daCosta *et al.* 1994) reveals similar features.

Full-sky redshift surveys allow a measurement of the gravitational acceleration acting on our local volume (e.g. Lynden-Bell *et al.* 1989). Recent analyses suggest that the source of our motion with respect to the microwave background extends beyond the nearest ~ 80 Mpc (Lauer and Postman 1994). The large swath cut out by the ZOA further complicates the interpretation of optical dipoles. Far-IR surveys reduce the ZOA, but they are sparse and are largely restricted to spiral galaxies, which may not fairly represent the mass (e.g. Strauss *et al.* 1992; Saunders *et al.* 1994). In order to fully understand the sources of motion nearby, we require dense redshift surveys in the ZOA.

GALACTIC PLANE SURVEY STATUS

In the spring of 1990, we began a redshift survey of the regions that are covered by the Zwicky-Nilson catalogue (Zwicky *et al.* 1961-68; Nilson 1973) but were not originally included in the CfA survey regions (the GPS). Our goals were to explore the relation between the Great Wall and the Perseus-Pisces Supercluster, and to begin what we hope will eventually be a complete all-sky survey of galaxies that samples the density field to redshifts near 0.05 (eg. Santiago *et al.* 1994). The nearly completed portions of the GPS now include the regions $4^h \leq \alpha \leq 8^h$, $0° \leq \delta \leq 90°$ and $17^h \leq \alpha \leq 20^h$, $0° \leq \delta \leq 90°$.

Galaxies in the GPS are drawn from the Zwicky Catalog. Although better photometry exists for a small fraction of these galaxies, we prefer to quote the original Zwicky magnitude in the interest of consistency. We quote the apparent magnitude in the B(0) system, which is indistinguishable from the Zwicky scale (Huchra 1976). We do not correct the magnitudes for extinction in the Galaxy or for internal extinction. A small fraction of the galaxies in our sample are multiple systems for which Zwicky *et al.* (1961-68) estimate only a combined magnitude. For these cases, we have estimated the relative contributions of each component by eye.

We obtained the vast majority of the GPS redshifts at the Tillinghast 1.5m telescope on Mt. Hopkins. We culled the remaining redshifts from the literature. A few of the faintest galaxies required observations at the Multiple Mirror Telescope. We measure redshifts by cross-correlating galaxy spectra with stellar and galactic templates (Tonry and Davis 1979) or by fitting Gaussian profiles to emission lines. Averaged over the sample, the mean error in cz is approximately 35 km s^{-1}.

Table 1 gives a status report on the GPS and the parent CfA survey. As of March 1994, we have completed nearly 89% of the CfA survey (nearly 14,000 galaxies) and 92% of the GPS (2455 galaxies out of 2667).

RESULTS

Figures 1 and 2 show the combined results of the GPS and the CfA surveys as well as the reshifts for other published data in the CfA Redshift Catalogue (Huchra *et al.* 1992). The data are presented as pie diagrams in redshift space. The most striking feature of the GPS is the wall enclosing the western void of CfA2N at $\approx$ 17-19^h and 8000-3000 km/s. The Great Wall is *not* simply connected to PP. Both the GW and PP form the boundaries of a network of voids with characteristic scale $\sim$ 50 h^{-1} Mpc. This western wall is not an artifact of absorption in the plane. We have checked this by comparing the peak predicted in the redshift distribution given our magnitude limit and the reddening map of

Burstein and Heiles (1982) to the observed redshift distribution. The observed wall does not follow the predicted peak and is significantly beyond it.

CONCLUSIONS

Combined with the CfA survey, the GPS confirms the view that the large structures we see, which appear to be coherent on scales of $\sim$ 100 Mpc, are merely the boundaries of several neighboring voids. The scale of these voids, 50 h^{-1} Mpc, is perhaps the most important physical scale for understanding the origin of structure.

We find that the GW and the PP chain are not simply connected across the zone of avoidance. The GW and PP are not entities unto themselves; they are formed at the intersections of neighboring voids.

REFERENCES

Burstein, D. and Heiles, C. 1982,AJ,87,1165

da Costa, L., Geller, M., Pellegrini, P., Latham, D., Fairall, A., Marzke, R., Willmer, C., Huchra, J., Calderon, J., Ramella, M. and Kurtz, M. 1994,ApJL

Geller, M. and Huchra, J. 1989,Science,246,897

Huchra, J. 1976,AJ81,952

Huchra, J., Geller, M., Clemens, C., Tokarz, S. and Michel, A. 1992,Bull. CDS,41,31

Lauer, T. and Postman, M. 1994, ApJ,425,418

Lynden-Bell, D., Lahav, O. and Burstein, D. 1989,MNRAS,241,325

Nilson, P. 1973, *Uppsala General Catalogue of Galaxies*, Uppsala Astron. Obs. Ann. 6.

Santiago, B., Strauss, M., Lahav, O., Davis, M., Dressler, A. and Huchra, J. 1994, ApJ, in press

Saunders, W., Sutherland, W., Efstahiou, G., Tadros, H., Maddox, S., McMahon, R., White, S., Rowan-Robinson, M., Oliver, S., Keeble, O., Frenk, C. and Smoker, J. 1994, this volume

Strauss, M., Yahil, A., Davis, M., Huchra, J. and Fisher, K. 1992, ApJ,397,395

Tonry, J. and Davis, M. 1979,AJ,84,1511

Zwicky, F. Herzog, E., Wild, P., Karpowicz, M. and Kowal, C. 1961-1968, *Catalogue of Galaxies and of Clusters of Galaxies*, (Pasadena: Caltech).

right ascension

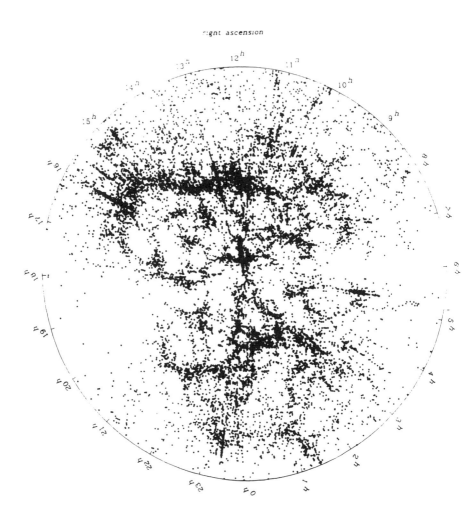

Figure 1.   The distribution of all measured redshifts in the 30 degree
declination wedge with $0° \leq \delta < 30°$, and within 15,000 km/sec.

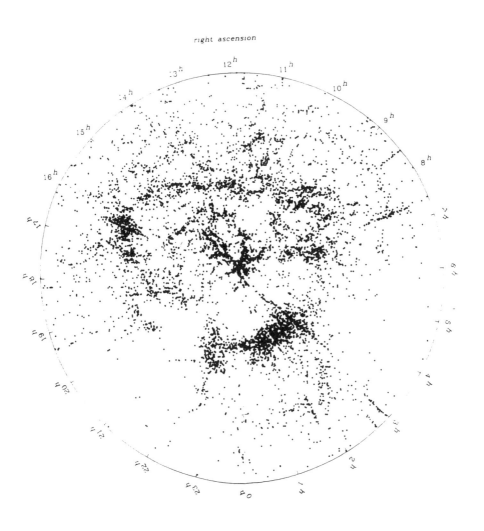

Figure 2.   As in figure 1 but for the 30 degree declination wedge with 30° $\leq \delta \leq 60°$.

Table 1 SURVEY ZONE SUMMARY   March 1, 1994

ZONE	α	δ	N_{gal}	N_{done}
NORTH GALACTIC CAP				
+0	8^h to 17^h	+00.0 to +02.5	538	294 **
+6	"	+02.5 to +08.5	1749	1218 **
+12	"	+08.5 to +14.5	1668	1667 comp
+18	"	+14.5 to +20.5	1365	1356 comp
+24	"	+20.5 to +26.5	1008	1008 comp
+30	"	+26.5 to +32.5	1136	1120 comp
+36	"	+32.5 to +38.5	725	725 comp
+42	"	+38.5 to +44.5	710	710 comp
+48	"	+44.5 to +50.5	570	495
+54	"	+50.5 to +56.5	599	308
+60	"	+56.5 to +62.5	426	237
Polar	"	+62.5 to +90.0	742	382
North Subtotals			11230	9520
SOUTH GALACTIC CAP				
+0	20^h to 4^h	-2.5 to +6.0	1105	1105 comp
+9	"	+6.0 to +12.0	587	587 comp
+15	"	+12.0 to +18.0	512	512 comp
+21	"	+18.0 to +24.0	411	411 comp
+27	"	+24.0 to +30.0	493	493 comp
+33	"	+30.0 to +36.0	624	624 comp
+39	"	+36.0 to +42.0	434	434 comp
+45	"	+42.0 to +48.0	188	188 comp
Polar	"	+48.0 to +90.0	103	101 comp
South Subtotals			4457	4455
CURRENT TOTALS			15698	13977
Plane Survey — to 15.5				
Summer	17^h to 20^h	+0.0 to +90.0	1200	1123
Winter	4^h to 8^h	" "	1467	1332

N_{gal} is the number of galaxies in the zone.
N_{done} is the number of galaxies with redshifts.
** are the strips currently being observed.

Discussion

R. Kraan-Korteweg: With your newly unveiled circular connections in the ZOA, the Great Wall looks much less prominent. It seems less one coherent enormous structure, but more a 'chance' connection of various cells.

J. Huchra: Yes. By definition, one of the reasons we wanted to probe deeper beyond the ends of the original survey was to see if, in fact, we could show that the 'Great Wall' is something causually connected or if it is the result of superpositions of other structures. It is looking more and more like the latter statement is true.

G. Mamon: How long is the 'Great Tube'? Is it more significant an underdense structure than the Great Wall is an overdense structure?

J. Huchra: We see it at essentially all declination ranges and as Tony Fairall said, it continues on through to the south. So, $\sim$ 120 degrees long at $\bar{v} \sim 2500$ km/s $\rightarrow \sim$ 50 Mpc h^{-1} (perhaps in deference to Donald Lynden-Bell, we should call this the 'hot dog' void).

D. Lynden-Bell: How thick would I have to make the walls of the voids if the galaxies in them are to make up for all those that might have been swept out of the voids.

J. Huchra: I am not sure exactly what you mean but in terms of filling factors - the 'walls' are over dense with respect to the mean density by about a factor of 5 or 50 and occupy $\sim$ 10-20% of the total volume - i.e. the voids have a 'filling' or 'emptying' factor of about 80-90%. The walls also contain about 80-90% of all the galaxies. You can work it out.

O. Lahav: How do galaxies of different morphological types trace the filaments in your map?

J. Huchra: Both early and late types trace the same structures except for the cores of rich clusters, where the early types dominate. The late types (and the lower luminosity galaxies) are somewhat fuzzier tracers and are slightly less good tracers than the earlier types and more luminous galaxies. This is from the work of one of our students, Mike Vogeley, now at John Hopkins, on structure as a function of luminosity.

A. Fairall: Just a comment. The number of redshifts that you show is absolutely breathtaking!

M. Hauschildt-Purves: You always use declination slices for cone diagrams. I find those somewhat confusing because it is not done in great circles, it is cone shaped as soon as you go away from the equator. Have you considered doing it in a different way.

J. Huchra: Oh yes! The best way to present this data is as Tony Fairall did using a 3-D video with rotating frames, but we are just lazy. I have used slices in RA, Dec, galactic latitude and supergalactic coordinates, but a 3-D workstation does it better.

Unknown: In what way does the galaxy luminosity function affect your 3D number density distributions?

J. Huchra: The luminosity function translates into what we usually call the selection function, i.e. the measure of how much of the galaxy luminosity function we can see as a function of distance. Nearby we see in our magnitude limited surveys even faint galaxies, at large velocities only the brightest. Because the luminosity function (in number/unit volume/magnitude interval) is nearly flat with an exponential cutoff at the bright end, we sample space moderately uniformly (slowly decreasing SF) out to ~ 8000 km/s. Then less well out to about 12,000 km/s. There are very few galaxies in our sample beyond 15,000.

Unveiling Large-Scale Structures Behind the Milky Way
ASP Conference Series, Vol. 67, 1994.
C. Balkowski and R. C. Kraan-Korteweg (eds.)

A Visual Search for Galaxies behind the Southern Milky Way on the UK Schmidt Atlas

Mamoru Saitō

Department of Astronomy, Faculty of Science, Kyoto University, Sakyo-ku, Kyoto 606-01, JAPAN

Abstract. This report reviews our visual search for galaxies behind the Milky Way region between $\ell \sim 210°$ and $250°$ at $|b| \leq 12°$ using the UK Schmidt Southern Infrared Atlas. About 7000 galaxies with diameters larger than 0.'1 were detected in about 900 $\deg^2$. The detection rate exponentially decreases with increasing H I column density and the characteristic column density is 3.3×10^{21} cm^{-2} for a 0.'2 size-limited search. The search identified a supercluster located at ~ 11000 km s^{-1} and a clustering of nearby galaxies.

1. Introduction

In order to study the structures of galaxy distribution in the Local Universe, we must reveal the galaxy distribution behind the Milky Way. The Milky Way has interfered optical systematic search for galaxies in this region. The Galactic extinction becomes less severe for longer wavelengths of observation. Since the UK Schmidt Southern Infrared Atlas on the Milky Way has completed in 1980's, we tried a systematic search for galaxies by means of this Atlas in 1988 and 1989.

2. The Search

We selected our servey region at the outer Galaxy where the Galactic extinction is known to be lowest among the zone along the Galactic plane. We surveyed two adjacent regions by two researcher groups, respectively (Saitō et al 1990, 1991a). Figure 1 shows the survey regions and the distribution of the detected galaxies; the dots are the detected galaxies and the contours indicate the H I intensities. The northern region is in the Monoceros constellation and the Galactic longitude is from about 210° to 230°. We call it region I. The southern region is in the Puppis at $\ell \sim 230°$ to 250°. We call it region II. These regions are just located on the $-SGZ$ direction of the Supergalactic coordinate. The H I column densities almost range of the order of 10^{21} cm^{-2}.

The procedure of our search is a visual inspection of the film copies of the Atlas using magnifier. Two researchers independently inspected the same field and identified the objects of galaxy-like image with diameter larger than 0.1 mm on the film copy; this critical size corresponds to 6.7 arcsec on the sky. The effective wavelength of the Atlas is 0.79 μm and the limiting surface brightness has been estimated to be 21.4 mag arcsec^{-2} (Hartley and Dawe 1981). The galaxies detected in this survey are brighter than about 17.5 in the I magnitude. After the first search the two researchers again examined the objects detected by either of them and decided its acceptance.

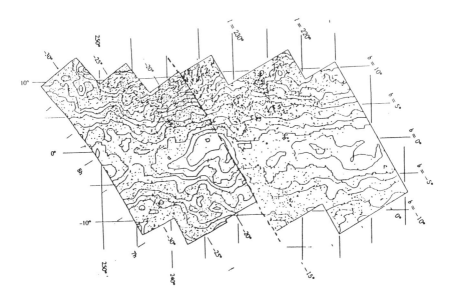

Fig. 1. The surveyed region and sky distribution of the detected galaxies (dots). The right-hand side of the dashed line is region I and the left region II. The contours indicate intensities of H I emission line (Bleomen 1983); the outermost line represents $N(HI) \sim 1 \times 10^{21}$ cm^{-2} and the innermost $\sim 1 \times 10^{22}$ cm^{-2}.

TABLE I The Visual Search for Galaxies behind the Southern Milky Way

Survey	Region I	Region II
Galactic longitude	$\sim210^{\circ}$ - 230°	$\sim230^{\circ}$ - 250°
Area (deg^2)	~500	~400
Material	UK Schmidt Southern Infrared Atlas on the Milky Way	
	18 fields	14 fields
Criterion		
the minimum diameter	0.1 mm	0.1 mm
Galaxy detected	2411	4633
Cataloged galaxy	31	194
Positional coincidence		
Radio source	16	10
IRAS point source	84	232
New detection	98%	96%

We made a cross identification of the selected objects by using the galaxy catalogs, *IRAS* Point Source Catalog, and catalogs of planetary nebulae, reflection nebulae, and other Galactic objects. If a selected object shows a positional coincidence with Galactic objects, we omitted it from our sample.

The positions of the detected galaxies were measured in a mean error of about 10 arcsec and were transformed to the equatorial coordinates by using SAO stars. All the detected galaxies are listed in two catalogs for region I and region II, respectively (Saitō et al. 1991b).

We summarize our search in Table I; 2411 galaxies were identified in about 500 deg^2 in region I and 4633 galaxies in about 400 deg^2 in region II. The rate of new detection is greater than 96%.

The galaxy number density in region I is about a half of that in region II. The difference is mainly caused by the difference of measurements of galaxy diameter between the two research groups. In an overlap zone of the two regions, the numbers of the detected galaxies can be compared with each other; Figure 2 shows the numbers as functions of galaxy diameter; the diameters measured by the second group are about 0.07 mm larger than those by the first group, and the resultant number of the detected galaxies becomes twice. Practically the first group adopted the mean values of the measurements by two researchers, while the second group adopted the larger values of the two measurements.

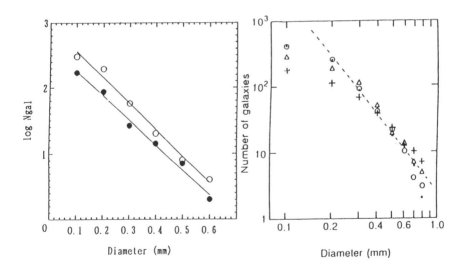

Fig. 2. (left) Size distribution of the detected galaxies in an overlap zone; the dots indicate galaxy numbers obtained by the research group of region I and the circles by the research group of region II.

Fig. 3. (right) Size distribution of the detected galaxies in four zones in region I. The dots and circles indicate zones with $N(HI) = 1 - 2 \times 10^{21}$ cm^{-2} and area 33 and 48 deg^2, the triangles a zone with $N(HI) = 3 - 4 \times 10^{21}$ cm^{-2} and area 75 deg^2, and the crosses a zone with $N(HI) = 7 - 9 \times 10^{21}$ cm^{-2} and area 147 deg^2.

The completeness of our detection can be evaluated from the following two data. Figure 3 shows the diameter distribution functions of the detected galaxies in the zones of different H I column densities; the spatially homogeneous distribution is shown by the dotted line, and deviations from the line occur for galaxies with diametrs smaller than 0.3 mm in the most regions. Secondly in the overlap zone of region I and region II, 91% of the galaxies with diameters equal to or larger than 0.3 mm in region I were detected in the search of region II. The rate becomes 96% for 0.4 mm. These mean that our detection is nearly complete for galaxies with apparent diameters equal to or larger than 0.3 mm at zones with H I column densities less than 7×10^{21} cm^{-2}. The critical diameter become 0.4 mm at the zones of higher H I column densities. We could not evaluate effects of the qualities of the film copies on the detection rate.

3.   Results

3.1 Detection Rate of Galaxies as a Function of H I Column Density
    Figure 4 shows the number densities of the detected galaxies in region I as a function of Galactic latitude; the number densities decrease toward latitudes closer to the Galactic plane, but the minimum is at $b \simeq -2°$. In Figure 4 we also show the mean H I column densities at two longitudes in region I (214.°5 and 224.°5); the H I distributions are somewhat asymmetric with respect to the Galactic plane and the peak is at $b \simeq -2°$.
    In region I the H I column densities almost parallely change with Galactic latitude, as shown in Figure 1. Thus the number densities of the detected galaxies are expressed as a function of H I column density. These are shown in the upper panel of Figure 5. The ordinate is logarithm of the number density of galaxies detected per square degree. The left panel indicates the number densities of the detected galaxies with diameters equal to or larger than 0.4 mm and the right panel is same as the left panel, but for the diameters equal to or larger than 0.2 mm. The regression lines in the left and right diagrams represent, respectively,

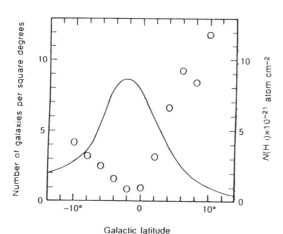

Fig. 4. The circles: Densities of galaxies detected in region I as a function of galactic latitude. The solid line: Mean H I column densities at $\ell = 214.°5$ and 224.°5.

$$N_{gal} = 1.4 \exp[-N(\text{H I})/4.8 \times 10^{21}] \ \text{deg}^{-2}$$

and

$$N_{gal} = 8.5 \exp[-N(\text{H I})/3.3 \times 10^{21}] \ \text{deg}^{-2},$$

that is, in our size-limited search, the number densities of the detected galaxies, N_{gal}, exponentially decrease with increasing H I column density, $N(\text{H I})$; the characteristic H I column density is 4.8×10^{21} cm^{-2} for the left panel and 3.3×10^{21} cm^{-2} for the left panel.

In region II, the H I column densities are not a simple function of Galactic latitude. Moreover, in some regions, the number densities of galaxies are distinctly higher than the surroundings. We divided region II into 70 areas and classified these areas in ten classes of the H I column densities. In the lower panel of Figure 5, the number densities of galaxies in region II are shown as functions of H I column density; in the diagrams, the dots refer to the high-density areas and the open circles the other areas. The galaxy densities similarly decrease with increasing H I column density to those in region I, although the densities are higher than in region I, even in the usual region other than the high-density region, mainly due to the difference of measuring procedure of galaxy sizes between the two regions. It is noticed that the difference of number densities of galaxies between the high-density areas and other areas is distinctly higher than the difference caused by the two measuring procedures.

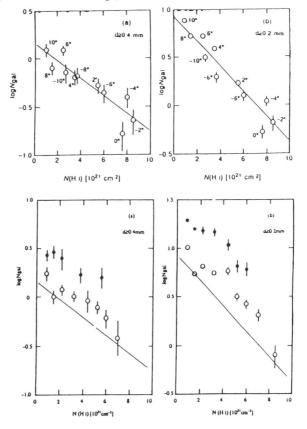

Fig. 5. Densities of the detected galaxies as a function of H I column density. The upper panels show the results for the two size-limited searches in region I and the lower panels the same but in region II. The lines in the lower panels are the regression lines in region I.

3.2 Decreasing of Apparent Diameters and Luminosities with H I Column Density

Figure 6 shows the diameter distribution functions of the detected galaxies in zones of the mean H I column densities of 1.0×10^{21}, 3.2×10^{21}, and 8.1×10^{21} cm^{-2} in region I, and 0.9×10^{21} and 5.6×10^{21} cm^{-2} in region II. The distributions yield a relation that the apparent diameters of galaxies exponentially decrease with increasing H I column density as

$$\theta/\theta_0 = \exp\left[-N(\text{H I})/1.37 \times 10^{22}\right].$$

The characteristic H I column density is 1.37×10^{22} cm^{-2}. The relation is consistent with the previously mentioned relation; i.e., in the size-limited search, the number densities of the detected galaxies exponentially decrease with increasing of H I column density.

From the emprical relation, we find that if the diameter is one arcmin at non-extinction region, the apparent diameter become 0.'92 at the region with H I column density of 1×10^{21} cm^{-2} and 0.'48 at the region of $N(\text{H I})= 1 \times 10^{22}$ cm^{-2}. At the regions with negligibly small Galactic extinction in this survey, the galaxy diameter of one arcmin corresponds to about 1.9 arcmin of the ESO survey on the *B*-band. Since our survey is nearly complete for galaxies with diameters larger than 0.'3, our survey is deeper than the ESO/Uppsala survey at the regions with H I column densities lower than about 7×10^{21} cm^{-2}. It should be noted, however, that in the visual search, we tend to preferentially detect galaxies located at the positions with relatively lower Galactic extinction within a $0.°5$ H I beam size. In fact, the characteristic H I column density 1.37×10^{22} cm^{-2} and the surface brightness distribution of galaxies yield an extinction coefficient 0.6 times as large as the usually accepted value at the *I*-band (Saitō et al. 1990).

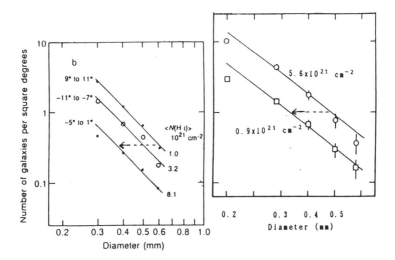

Fig. 6. Diameter distribution functions of the detected galaxies in zones of different mean H I column densities. The left diagram is for region I and the right for region II.

Galaxies become less bright with increasing H I column density. Saitō, Takata, and Yamada (1994) compared the blue magnitudes of IRAS/UGC and IRAS/CGCG galaxies at the Milky Way region ($|b| < 15°$) with those at $b = 30°$ to 45°. Figure 7 and Table 2 show the results. Blue luminosity decrements due to extinction were estimated by Cameron (1990) for spiral galaxies. The luminosity decrements in I-band are less steep than in the B-band. Saitō et al. (1994) evaluated the depth of a visual search for galaxies behind the Milky Way; for example, if we detect galaxies with apparent magnitude up to $m_B = 20$ or $m_I = 17.5$ at the region with H I column densities less than 6×10^{21} cm^{-2}, we see the galaxies up to 15.5 mag in B and I bands in the clear region.

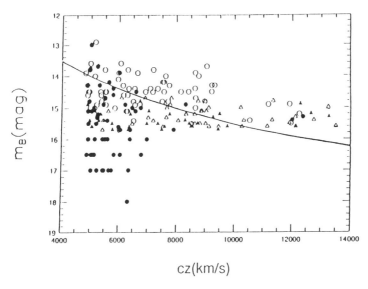

cz(km/s)

Fig. 7. The blue magnitudes of IRAS/UGC (circles) and IRAS/CGCG (triangles) galaxies versus the redshifts. The filled and open symbols indicate galaxies at $|b| < 15°$ and $b = 30°$ to 45°, respectively. Their galactic longitudes are from 150° to 242°. The curve indicates $M_B = -19.5 + 5\log(H_0/100)$.

TABLE II Magnitude Distributions of the Galaxies Shown in Figure 7 with Redshifts of 4800 to 6500 km s^{-1} for Six Latitude Ranges

Magnitude			Galactic	latitude		
	30° to 45°	13° to 15°	11° to 13°	9° to 11°	7° to 9°	3° to 7°
Total number	38	19	18	10	8	8
$m_B \leq 15$	30	10	4	3	0	0
$16 > m_B > 15$	8	8	9	1	4	2
$m_B \geq 16$	0	1	5	6	4	6
Mean	14.6±0.6	14.9±0.8	15.5±0.7	15.8±1.0	16.1±0.5	16.4±0.8

3.3 Clustering of Galaxies in the Surveyed Region

In our surveyed regions, we have found two clusterings of galaxies. In region I there are three zones with higher densities of galaxies around $b = 5°$. We measured redshifts of almost all the IRAS galaxies and the brightest non-IRAS galaxies existing in and around the three zones. The three clusters are nearly the same distance around redshifts of 11000 km s^{-1} and form a supercluster (Yamada and Saitō 1993). We called it the Monoceros supercluster.

In front of this supercluster, a less dense region extends over redshifts of 2000 to 8000 km s^{-1}. The spatial extent of the less dense region can be seen in the IRAS galaxy distribution made by Takata et al. (1994).

In region II a filamentary clustering of large nearby galaxies was found at around $\ell = 240°$ and $b = -12°$ to 10°. The clustering of galaxies in the Puppis has been studied by Yamada et al. and the detail is reported in this workshop by Yamada.

Acknowledgments. I thank the Yamada Science Foundation for providing the financial support for my participation to this workshop.

References

Bleomen, J. B. G. M. 1983, in Survey of the Southern Galaxy, ed. W. B. Burton and F. P. Israel (Dordrecht: D. Reidel Publishing Company), 307

Cameron, L. M. 1990, A&A, 233, 16

Hartley, M., Dawe, J. A. 1981, Proc. Astron. Soc. Australia, 4, 251

Saitō, M., Ohtani, H., Asonuma, A., Kashikawa, N., Maki, T., Nishida, S., Watanabe, T. 1990, PASJ, 42, 603

Saitō, M., Ohtani, H., Baba, A., Hotta, N., Kameno, S., Kurosu, S.,Nakada, K., Takata, T. 1991a, PASJ, 43, 449

Saitō, M., et al. 1991b, Catalog of Galaxies behind the Milky Way, Vol. 1 & Vol. 2, Contribution Dept. Astron. Kyoto Univ. No. 300

Saitō, M., Takata, T., Yamada, T. 1994, PASJ, 46, in press

Takata, T., Yamada, T., Saitō, M., Chamaraux, P., Kazés, I. 1994, A&AS, 104, in press

Yamada, T., Saitō, M. 1993, PASJ, 45, 25

Discussion

R. Kraan-Korteweg: What is the slope of your linear relation in your figure of log D (mm) versus log N? Does it differ from -3, the slope for unobscured regions?

M. Saitō: The total number of galaxies detected in an area increases with the cube of the limiting size, when the spatial distribution of galaxies is homogeneous. Our survey shows that the relation is approximately realized even behind the Milky Way region, if the area considered contains 200 or more galaxies and has similar HI column densities. The large deviation from this relation occurs near the limiting size in our search. This is caused by the incompleteness of the search and the critical size may be an indicator of the completeness limit of the search.

E. Meurs: In one of your first viewgraphs, you showed a table, in which numbers of positional coincidences with radio and IRAS sources were given. For the IRAS sources, does this refer to purely positional coincidence or did you also check whether their IRAS colours are consistent with being galaxies?

M. Saitō: Yes. We made the IRAS colour-colour diagrams of the selected objects associated with IRAS PS, and found that the infrared colours of such objects are consistent with those of galaxies. Using the colour-colour diagrams, Yamada *et al.* have optimized the infrared criteria for selecting IRAS galaxies behind the Milky Way.

C. Balkowski: Where did you measure the redshifts of the IRAS galaxies of your sample?

M. Saitō: We carried out the redshift survey at the Okayama Astrophysical Observatory using the 1.88-m reflector.

Unveiling Large-Scale Structures Behind the Milky Way
ASP Conference Series, Vol. 67, 1994.
C. Balkowski and R. C. Kraan-Korteweg (eds.)

Galaxy Surveys Near the Galactic Center Region and the Ophiuchus Supercluster

Ken-ichi Wakamatsu[1]

Physics Department, Gifu University, Gifu 501-11, Japan

Takashi Hasegawa[2]

Department of Astronomy, The university of Tokyo, Tokyo 113, Japan

Hiroshi Karoji

National Astronomical Observatory, Mitaka 181, Japan

Kazuhiro Sekiguchi and John. W. Menzies

South African Astronomical Observatory, P O Box 9, Observatory 7935, Cape, South Africa

Matthew Malkan[1]

Department of Astronomy, UCLA, Los Angeles, CA 90024, U.S.A.

Abstract. Extensive galaxy- and redshift-surveys have been carried out around the Ophiuchus cD cluster which was discovered at l =0.5°, b = +9.5° by Wakamatsu and Malkan (1981) as the 3rd brightest X-ray cluster in the sky. The cluster is found to be accompanied by several irregular clusters and so to form the "Ophiuchus supercluster." This supercluster may be connected with the Hercules supercluster by a wall structure.

1. Introduction

The galactic center region is one of the most difficult and so the most challenging areas for galaxy surveys due to a high density of complex dust clouds, emission and reflection nebulae, and foreground stars. Nobody would undertake a galaxy survey there without the strong motivation of investigating a nearby large scale structure of galaxies.

During his surveys of variable and proper motion stars on Palomar I-N Schmidt plates near the Galactic center region, Terzan (1966, 1971) serendipitously detected several tens of fuzzy objects within a distance 10° from the

---

[1]Visiting Astronomer, Cerro Tololo Inter-American Observatory. CTIO is operated by AURA, Inc. under cooperative agreement with the National Science Foundation

[2]Visiting Astronomer, South African Astronomical Observatory

Galactic center. Though a few objects turned out to be globular clusters (Terzan et al. 1988), e.g., Trz 2, 3 and 4, most of them were believed to be nuclei of heavily obscured galaxies, and subsequently confirmed as such by Djorgovski et al. (1990), and in the present redshift survey. During our globular cluster survey on deep Palomar Schmidt IV-N plates, Wakamatsu and Malkan (1981) discovered a rich cD cluster at the position l =0.5°, b =+9.5°, just 1.5° north of the Terzan survey area. This cD cluster has been called the Ophiuchus cluster, and turned out to be the third brightest X-ray cluster in the sky after the Virgo and Perseus clusters (Johnston et al. 1981, Böhringer et al. 1991, Kafuku et al. 1991). The cluster also has several strong radio sources, e.g., MSH 17–203 and MSH 17–205 (Wakamatsu and Malkan 1981). The Ophiuchus cluster has a recession velocity cz = 8500 ± 150 km/s, and so lies at a distance comparable to the Coma cluster.

Encouraged by this discovery, we have been conducting extensive galaxy- and redshift-surveys around the Ophiuchus cluster for the following purposes:

1. Whether there are nearby rich clusters that could affect the bulk motion of the Local Group of galaxies,

2. Whether the Ophiuchus cluster is associated with a few rich clusters, and so forms the "Ophiuchus supercluster",

3. If so, whether there are any wall structures connecting Ophiuchus and other nearby superclusters, such as the Hercules, Hydra-Centaurus, and/or Pavo-Indus-Telescopium superclusters.

In this paper, we present our preliminary results, and details will be published in a forthcoming paper (Hasegawa et al. 1994) after the data analyses are completed.

2. Galaxy Survey

Galaxy surveys were made in two series: (1) a narrow but *deep* survey in order to examine the presence of a supercluster around the Ophiuchus cD cluster and (2) a wide but *shallow* survey in order to examine the presence of wall structures around the Ophiuchus cD cluster. The former survey covers the 6 SERC fields #453, 454, 518, 519, 586, and 587, while the latter covers the region 16h 10m < RA < 17h 50m and −32.5°< Dec < 0°.

We used ESO/SERC Sky Survey films of J, B, and I (when available) bands. The films were mounted on a twin-stage x-y measuring machine, and inspected with Nikon binocular-type microscopes (magnification ×15). Both images of J and I or B films of the same field mounted on the twin-stage can be displayed simultaneously on TV screens through CCD television cameras attached to the two microscopes. These images are very useful for comparing image qualities on the different color bands and for measuring angular sizes of detected galaxies.

We tried to pick out images that appear non-stellar with angular diameter larger than 0.04 mm (2.7″) and 0.1mm (6.7″) for deep and shallow surveys, respectively. Completeness of the surveys becomes worse for galaxies with smaller sizes. Many galaxies are superposed on foreground stars, while others appear barely distinguishable from stellar objects because only bright nuclear regions can be seen for highly obscured galaxies. For fields #453 and 519, where both J and I films are available, most of the galaxies appear more clearly on J films than on I films, and unexpectedly there exists no object that can be detected only on I films. This implies that most of this area is either transparent enough to see beyond the Milky Way even in the J band, or opaque even in the I band. We found no advantage to making a galaxy survey in the I band.

We detected 3900 objects in total for the deep survey, and measured their positions within an accuracy of $\pm 3''$, magnitude classes (MC), angular diameters, and morphological types. Luminosity of the objects is assigned into five steps, MC 1 ($\sim m_J = 15.5 \pm 0.5$) to MC 5 ($\sim m_J = 19.5 \pm 0.5$), with intervals of 1 magnitude. Due to large differences of foreground star densities, errors may amount to 1 magnitude for some objects. Detected galaxies are classified into elliptical, S0, spiral, barred spiral, compact (semi-stellar), and diffuse. Most of the objects classified as compact or semi-stellar may be bright nuclei of highly obscured E or S0 galaxies. Objects with low surface brightnesses without cores are classified as diffuse. They may be Magellanic-type dwarf galaxies, but some might be galactic objects. Their lists and finding charts will be published in the future.

3. Redshift Survey

Slit spectra have been obtained with the CTIO 1.5m and SAAO 1.9m telescopes for bright galaxies of MC 1 and 2, and with the Lick 3m telescope for faint galaxies of MC 3. At CTIO, we started spectroscopic observations in 1983 and 1984 by using the Cassegrain SIT vidicon spectrograph, but it took 1.5 hours to obtain a good enough spectrum to measure the recession velocity. This is because the blue spectral region where the SIT vidicon had high sensitivity is heavily absorbed in our program galaxies. In 1992 and 1993, after a CCD detector was installed, good spectra in the red region were easily obtained with a typical exposure time of only 10 minutes, and the efficiency of our redshift survey improved greatly. CCD spectra were obtained with a dispersion of 4.21 Å/pix for a spectral range of 4806 Å $< \lambda <$ 7220 Å. At SAAO, we used a Cassegrain reticon spectrograph both in 1992 and 1993 to cover a wavelength range of 3400 Å – 7400 Å with 2.8 Å/pix. At Lick, the KAST double-spectrograph at the Cassegrain focus was used in 1993 for those objects which are faint but large and so plausibly foreground galaxies. The red channel gave spectra with 1.70 Å/pix for a range of 5467 Å – 7490 Å; the blue channel gave very weakly exposed spectra.

Spectra of 300 objects have been obtained so far, and recession velocities of 280 objects can be measured in total, with a typical accuracy of ± 70 km/s. For the deep survey region, 147 galaxies were measured, comprising 87/128=68%, 39/374=10%, and 15/774=2.4% of the MC 1, 2, and 3 galaxies, respectively. They were selected evenly in the survey area, but for concentrated regions they were picked out so as to check whether these concentrations form physical systems or not. Note that this redshift survey is not complete in any sense.

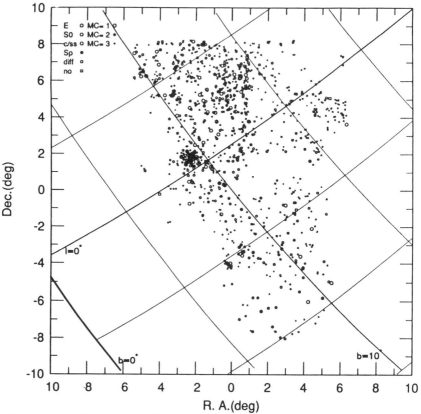

Figure 1.   The spatial distribution of galaxies of MC 1 – 3.   The center of the display is at RA = 17h 00m, Dec = -25° 00′.

For the shallow survey region, a very sparse redshift-survey has been done for only 11 SERC fields so far, at a rate of 4 – 6 galaxies per field, and redshifts have been measured for only 60 galaxies.

4. Ophiuchus Supercluster

The spatial distribution of galaxies of MC 1 – 3 is shown in Figure 1 for the deep survey region. Curves of Galactic longitude and latitude are drawn with intervals of 5°. Galaxies were detected for the region $b > +6°$ only. An empty zone running E-W at Dec = $-24.0°$ with a width of $\sim 1.5°$ is the famous heavy dust lane starting from the ρ Ophiuchus cloud towards the east. Some of the empty zones in Figure 1, having angular sizes larger than 1° × 1° correspond to prominent dust clouds that were easily noticed on SERC J films due to their deficiency of foreground stars.

A prominent concentration in Figure 1 at RA = 17h 09m, Dec = $-23°$ 18′ is the Ophiuchus cluster, which looks very rich enough to be classified as a cD

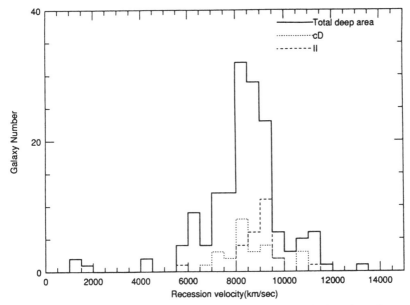

Figure 2.   Redshift histograms of galaxies for the Ophiuchus cluster
region.

cluster. A redshift histogram of galaxies lying within 1.5° of the cluster center is
shown as a dotted line in Figure 2. The cluster is found to have a mean recession
velocity of 8500±150 km/s and a large velocity dispersion of ∼1000 km/s, which
is consistent with its high X-ray luminosity (Okumura et al. 1988).

Another fairly prominent concentration in Figure 1 can be recognized at RA
= 16h 59m, Dec = −19° 20′ with a diameter ∼ 3°. It is enhanced by the presence
of a few empty areas around it. These are, however, not caused accidentally by
foreground dust clouds but a real enhancement, because most of these empty
areas are filled up with faint galaxies of MC 4 and 5, and so transparent enough
to detect bright galaxies, if they exist. This concentration is spiral–rich, and has
no prominent core, and so should be classified as an irregular cluster. Indeed,
as can be seen from its recession velocity histogram, drawn in a broken line in
Figure 2, for galaxies lying within 2° of its cluster center, it has a mean recession
velocity of 9000 ± 150 km/s and a small velocity dispersion ∼ 500 km/s. This
cluster is hereafter referred to as "Ophiuchus II."

Besides the Ophiuchus cD and II clusters, a few other real concentrations
can be recognized in this survey area, e.g., at (RA, Dec) = (17h 12m, −19° 53′)
and (16h 59m, −28° 44′), though they are not so conspicuous as the above two
rich clusters. These weak enhancements may be partly confused by patchy dust
clouds. Our present redshift survey indicates that they are physical systems,
and rich enough to be classified as clusters or groups of galaxies. Their mean
recession velocities are found to be also close to 8500 km/s.

A recession velocity histogram of all the galaxies measured in this deep
survey area is drawn as a solid line in Figure 2. Many of the bright galaxies

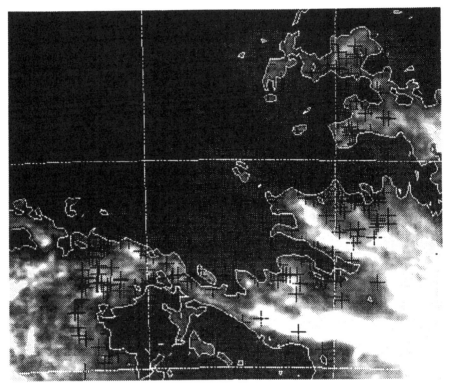

Figure 3.   A distribution of MC 1 – 3 Galaxies around the ρ Ophiuchus region is plotted on a grey scale 100 μm map from the IRAS Sky
Survey Atlas.

lying outside of the cluster areas are also found to have recession velocities in the
range 7000 km/s $< cz <$ 10000 km/s, implying that they lie at similar distances
as the above clusters. Therefore, this ensemble of galaxies in the survey area is
large and rich enough to be identified as a supercluster of galaxies. It has an
angular extent of at least 18° × 12°, and so a linear extent $27h^{-1}$ Mpc × $18h^{-1}$
Mpc, where $h = H/100$ km/s/Mpc. We call it the "Ophiuchus supercluster."
    Note that there are small groups at 6500 km/s and 11500 km/s in Figure 2.
The latter group is the cluster originally identified by Djorgovski et al. (1990)
at 1.8° south of the Ophiuchus cD cluster. It is barely recognizable in Figure 1,
and is not as rich as the Ophiuchus II cluster, though it is a real cluster. It is
not yet clear whether it belongs to the Ophiuchus supercluster or is just in the
background.

5.   How Heavily Is the Supercluster Region Obscured?

A fairly tight correlation is suggested between the color excess and IRAS 100
μm flux, $F(100)$, of the diffuse background (Boulanger and Perault 1988). In

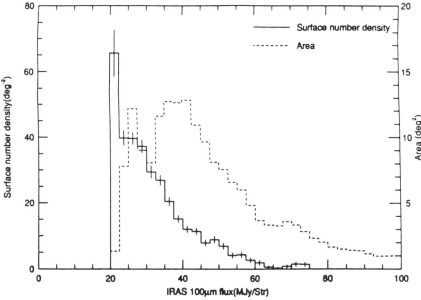

Figure 4.   A surface number density of galaxies (left-side scale) and total areas binned by their background IRAS fluxes (right-side scale) are plotted as a function of $F(100)$.

Figure 3, galaxies of MC 1 – 3 around the ρ Ophiuchus region are plotted on a rey scale 100 μm map from the IRAS Sky Survey Atlas, which was released in 993 and corrected for the zodiacal light. Black or white areas correspond to ones ith $F(100) < 30$ MJy/str or $F(100) > 80$ MJy/str, respectively. Generally, alaxies are found to avoid areas of high IRAS flux. Some empty zones are cated in low IRAS-flux areas, while the remaining empty zones lie around high RAS-flux areas. The former empty areas are found to be due to non-uniformity r clumpiness of the intrinsic galaxy distribution, as discussed below.

In Figure 4, a surface number density of detected galaxies of MC 4 – 5 is lotted as a function of their background IRAS flux density, $F(100)$ (solid line). ertical bars represent ambiguities coming from statistical errors $\sqrt{N\,gal}$, where gal is the total number of detected galaxies in each area binned by IRAS flux. he dashed line shows the total areas binned by IRAS 100 μm flux. As can e seen from Figure 4, the minimum flux in the present deep survey area is bout 20 MJy/str, while the maximum flux among areas where galaxies were etected is ~75 MJy/str. The curve of surface number density in Figure 4 is asonably smooth and decreases rapidly as IRAS flux increases. Since galaxies f MC=4 and 5 are not prominently clustered, the curve can basically represent le behavior of dust extinction as a function of $F(100)$. Thus the 100 μm sky flux an be a very useful estimator of Galactic absorption with an angular resolution f 1.5' which is higher than that of typical HI column density measurements.

Our CCD spectra of E and S0 galaxies are good enough to estimate a color cess of each galaxy by comparison with a standard galaxy spectrum. As a

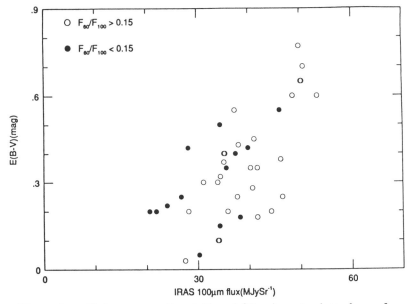

Figure 5.   Color excess of each E and S0 galaxy is plotted as a function of $F(100)$ at each galaxy position.

template we adopted the E2 spectrum given by Bica (1988). The resultant color excesses of each E and S0 galaxy are plotted in Figure 5, as a function of $F(100)$ at each galaxy position, with two different symbols for warm dust regions $(F(60)/F(100) > 0.15)$ and cool dust regions $(F(60)/F(100) < 0.15)$. According to this preliminary result, there is a clear correlation but no strong dependency on dust temperature is found. Large scatters in the diagram may be partly due to intrinsic differences of an energy spectrum of each galaxy. The regression line is

$$E(B - V) = 0.022F(100) - 0.45.$$

This equation was obtained allowing for an offset of $F(100)$ at $E(B - V) = 0$. However, it is not yet clear whether this term is really necessary or not. By using the regression line, the maximum color excess for the detected region can be estimated to be $E(B - V) = 1.2 \pm 0.2$ for the region with $F(100) = 75$ MJy/str. If we adopt $Q = 2.97$, this corresponds to $A_V = 3.6 \pm 0.6$, and seems reasonable. Consequently, galaxies can be detected on SERC J films only for areas obscured by less than $A_V \sim 3.6$ magnitude.

Radial variations of azimuthally averaged $F(100)$ for each annulus zone are examined as a distance of cluster center for the Ophiuchus cD and II clusters. They are nearly constant at 53 ± 5 MJy/str and 26 ± 3 MJy/str for the former and latter cluster, respectively, within a radius $r < 1.5°$. Based on these analyses, the total amount of color excesses are estimated to be $E(B - V) = 0.72 \pm 0.1$ and 0.12 ± 0.05 for the cD and II clusters, respectively. We can now examine spatial variations of dust obscuration by using IRAS 100 μm flux maps in this

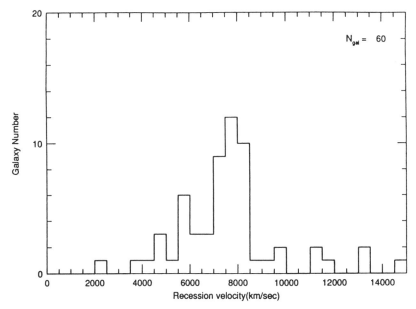

Figure 6.   A redshift histogram of galaxies for the shallow survey region.

region, and will continue a much more detailed study of the structure of each cluster.

6. Large Scale Structures and the Ophiuchus-Hercules Wall

For the shallow survey region, fairly wide areas remain highly obscured due to the heavy dust clouds belonging to the famous north polar spur, though this zone reaches galactic latitude $b = 30°$. Our galaxy survey for this wide area is still in progress, and no maps of spatial distribution of galaxies can be shown in the present paper. However, a preliminary redshift histogram is shown for all the measured galaxies in this zone excluding galaxies in the deep survey area (Figure 6).

From this histogram, the following features can be clearly noticed: (1) A peak at 8000 km/s is again the most prominent group. (2) A new peak appears at 4750 km/s. (3) The peaks recognized in Figure 2 at 6500 and 11500 km/s still remain. (4) The number of galaxies with $cz < 4000$ km/s is very small. (5) A reasonable number of background galaxies are detected, as expected. The deficiency of nearby galaxies listed in item (4) suggests the presence of a prominent local void in this area. We can tentatively conclude that, in this sky region, there are no nearby rich clusters of galaxies that could strongly affect the bulk motion of the Local Group of Galaxies.

Galaxies belonging to the 8500 km/s component are distributed all over the shallow survey zone, and appear to form a big ensemble. Indeed, many galaxies of similar appearances to those of galaxies with 8500 km/s are found in this survey region, and so most of them are probably physical members of the ensemble. In order to examine whether this ensemble extends just beyond our survey area to the north, we refer to the results of a redshift survey in the northern hemisphere. This is the Hercules region. Fortunately, Freudling et al. (1991) made extensive redshift surveys for the region 15h 30m < RA < 16m 30m, 0°< Dec < +60°, and obtained cone diagrams. On their Figure 8, a narrow but distinct plume having cz = 9000 km/s can be noticed starting from Dec = 11° towards the south, and is a southern extension of the Hercules supercluster. This plume may connect with the present big ensemble at 8500 km/s. If so, our ensemble and Freudling's plume may form the "Ophiuchus-Hercules Wall," starting at Abell 2199, and passing through Abell 2151, and ending at the Ophiuchus cD cluster. Its angular and linear extents are about 65° and $96h^{-1}$ Mpc, respectively.

Since the supergalactic coordinates of the Ophiuchus cD cluster are SGL = 175° and SGB = +45°, this wall lies at $60h^{-1}$ Mpc above the Local Supergalactic plane and runs parallel to it. This proposed wall crosses perpendicularly the Great Wall at the Abell 2199 cluster. Tully et al. (1992) suggested that wall structures have a tendency to cross each other perpendicularly. The present case may be a nearby example.

In order to establish the proposed wall and to examine a new wall structure towards Hydra-Centaurus supercluster, further redshift surveys are scheduled using the FLAIR multi-fiber spectrograph attached to the 122-cm UK Schmidt telescope.

Acknowledgments. We are grateful to Dr. Robert Williams, the former director of CTIO and Dr. Bob. Stobie, the Director of SAAO, for allocating observing time for this project. We acknowledge Dr. Mario Hamuy at CTIO for his hearty support for data reduction.

References

Bica, E. 1988, A&A, 195, 76

Böhringer, H. et al. 1991, in Frontiers of X-Ray Astronomy, eds. Y. Tanaka & K. Koyama, Tokyo: Universal Academy Press, 459

Boulanger, F., & Perault, M. 1988, ApJ, 330, 964

Djorgovski, S., et al. 1990, AJ, 100, 599

Freudling, W., Martel, H. & Haynes, M. 1991, ApJ, 377, 349

Hasegawa, T. et al. 1994, in preparation

Johnston, M.D. et al. 1981, ApJ, 245, 799

Kafuku, S. et al. 1991, in Frontiers of X-Ray Astronomy, eds. Y. Tanaka & K. Koyama, Tokyo: Universal Academy Press, 483

Okumura, Y. et al. 1988, PASJ, 40, 639

Terzan, A. 1966, *C. r. Acad. Sci. Paris* 263, Series B, 221

Terzan, A. 1971, A&A, 12, 477

Terzan, A. et al. 1988, A&AS, 76, 205

Tully, B. et al. 1992, ApJ, 388, 9

Wakamatsu, K. & Malkan, M. 1981, PASJ, 33, 57

Discussion

W. Saunders: Do the 100 μm sky fluxes you quote have the zodiacal light subtracted?

K. Wakamatsu: Yes they do.

L. Gouguenheim: Is there a velocity difference between the two clusters?

K. Wakamatsu: It amounts to less than 200 km/s.

E. Valentijn: The Hercules supercluster is the only structure for which there is strong evidence of a supercluster pervading intra/inter cluster medium. The evidence originates from the existence of head-tail radio sources located at the outskirts of the Abell clusters in Hercules (Valentijn, 1978). The total mass involved in the gas is very high and even more than seen in the galaxies. Your suggestion of a connection between the gas-rich Ophiuchus cluster and the Hercules supercluster seems to point at a very substantial large-scale gas-mass into that direction.

K. Wakamatsu: According to radio maps for a few strong radio sources, they are not point sources but extended sources.

A. Fairall: You mention that your new wall is perpendicular to the Great Wall. This, then, seems to be yet another example of walls intersecting at right angles.

K. Wakamatsu: We want to obtain strong evidence for the presence of the Ophiuchus-Hercules wall with the coming FLAIR observing run at the UK Schmidt telescope.

C. Balkowski: What uncertainty on the velocity do you expect from the FLAIR measurements?

K. Wakamatsu: We expect about ±100 km/s.

A Search for Galaxies in the Anticenter ZOA

C.A. Pantoja

Department of Physics and Astronomy, University of Oklahoma, Norman, OK 73019-0225

D.R. Altschuler

National Astronomy and Ionosphere Center - Arecibo Observatory [1], P.O. Box 995, Arecibo, PR 00613

C. Giovanardi

Osservatorio Astrofisico di Arcetri, Largo Enrico Fermi 5, I-50125 Firenze, Italy

R. Giovanelli

National Astronomy and Ionosphere Center and Department of Astronomy, Cornell University, 512 Space Sciences Building, Ithaca, NY 14853-6801

Abstract. We report on the current status of an HI survey to study the distribution of galaxies in the Anticenter region of the Zone of Avoidance. We have completed a catalog of galaxies by visual inspection of the POSS E prints in this region and observed the 21cm line of HI for a sample of galaxy candidates from this catalog to confirm their identification as galaxies and obtain redshifts.

1. Introduction

We report here on a project to map the distribution of galaxies in the region between $4h \leq \alpha \leq 8h$ and $0° \leq \delta \leq 37°$ (corresponding approximately to $160° \leq l \leq 220°$ and $-10 \leq b \leq 10°$). This is a region that is adjacent to the Lynx-Ursa Major supercluster (Giovanelli & Haynes 1982) on one side of the galactic plane and to the southwestern spurs of the Pisces-Perseus complex on the other (Giovanelli *et al.* 1986). It has been suggested by Giovanelli & Haynes (1982) and by Dow *et al.* (1988) that these structures might be connected across the Zone of Avoidance (ZOA). The connectivity of large-scale structures across the ZOA has important consequences for theories of galaxy formation, since any clear connections will significantly increase the extent of the largest coherent

---

[1]The Arecibo Observatory is part of the National Astronomy and Ionosphere Center. The NAIC is operated by Cornell University under a cooperative agreement with the National Science Foundation.

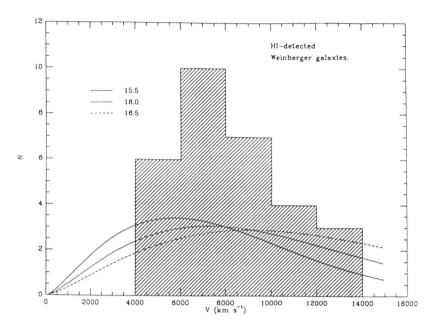

Figure 1.   The Heliocentric Velocity Distribution for the Weinberger
sample

structures observed. We have chosen a window in the galactic plane in which
extinction is low as compared to nearby galactic longitudes. We selected the
declination range to correspond to that accesible to the Arecibo Observatory
305m telescope to use the 21cm line of HI for observation of galaxy candidates
in order to confirm their extragalactic nature and at the same time determine
their redshifts. After a brief description of a pilot program, the rest of the paper
provides a description of the optical catalog and a discussion of results of the HI
measurements.

2.   The Study of Weinberger Galaxies

We initiated our study with a program to obtain HI spectra of a sample of low
galactic latitude galaxy candidates. Our sample was taken from a list compiled
by Weinberger (1980). Weinberger performed an optical search for galaxies in
the region $33° \leq l \leq 213°$ and $-2° \leq b \leq 2°$ by examination of the POSS E
prints with a 16x binocular microscope. His list contains 207 galaxy candidates
from which we selected 86 which lie within the Arecibo declination range ($0° \leq \delta \leq 37°$) to be observed in HI.

    We detected 30 of the 86 candidates observed ($\approx 35\%$).The distribution in
velocity is shown in figure 1. The superimposed curves on the histogram represent the expected distribution for a randomly distributed sample computed
using a Schecter luminosity function (with three different limiting apparent mag-

nitudes, ($H_0 = 50 kms^{-1} Mpc^{-1}, \alpha = 1.2, M^* = -1.2$). We see from this figure that the velocities peak at about $7000 kms^{-1}$.
For details of this study the reader is referred to Pantoja et al. 1994.

3. Optical Catalog

We completed a catalog of galaxy candidates by visual inspection of 29 POSS E prints in the region 4h $\leq \alpha \leq$ 8h and $0° \leq \delta \leq 37°$. For the search we used a setup in which the prints were illuminated from below and scanned with a 12× ocular. Each print was systematically scanned along horizontal lines. Images which appeared diffuse and were larger than 0.1mm were selected. When a candidate was found it was marked on a transparency that was placed over the print. Our objective was to find a reasonable number of galaxy candidates in each print so as to be able to obtain redshifts for a study of structure in the anticenter ZOA. After selection of the candidate on the print, the sizes were measured and an assessment of the quality of the image was made. The coordinates were measured using a Digitizing Mat interfaced to a PC running the program FINDER (Herter 1992). Positional errors are of the order 10″. The search resulted in a catalog of 1480 galaxy candidates (Pantoja et al. 1994).

We used the Atlas of Sky Overlay Maps for the Palomar Sky Survey (Dixon 1981) to identify cataloged galaxies or galactic objects. We used NED[1] to further search for identifications with previously cataloged objects (within a search radius of 3′). We found 92 of our candidates to correspond with previously cataloged galaxies of which 17 have measured redshifts. Of these 92 identifications, 25 were identified as proceeding from Weinberger's list (1980). The highest detection rate found in our pilot program (refer to section 2) was for diameters in the range 0.3′ $\leq$ d $\leq$ 0.5′ .Twenty of the objects (20/25) found using NED and associated with Weinberger galaxies have sizes larger than 3′ (9 objects of sizes larger than 3′ from Weinberger's list were not included in the catalog). We found that 104 objects could be associated with sources from the IRAS PSC (1988). Thirteen of these 104 objects have measured redshifts. In addition 74 objects were identified as radio sources.

In figure 2 we plot the distribution in galactic coordinates of the 1480 objects in the catalog. The solid line delineates the surveyed region. At lower latitudes it becomes more difficult to identify galaxies, a trend that has given the region its name. In the area around b=0° there is a large amount of obscuration due to absorption nebulae, reflection nebulae and HII regions. In particular around l=190° and l=200° there are two molecular clouds, Gemini OB1 and the Monoceros complex. The regions at b=-10° l=205° and l=170° are adjacent to the Orion complex and the Taurus clouds. We find almost twice as many galaxy candidates above the galactic plane than below (64% above and 36% below). If most of the candidates are actually galactic objects, rather than galaxies, then this asymmetry is understood, because the Galactic plane is warped towards $b > 0$ in the anticenter. There is a concentration of objects above the plane be-

---

[1]"The NASA/IPAC Extragalactic Database (NED) is operated by the Jet Propulsion Laboratory, California Institute of Technology, under contract with the National Aeronautics and Space Administration."

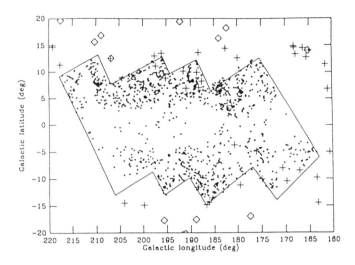

Figure 2.   Optical Catalog. The 1480 galaxy candidates plotted in galactic coordinates. The solid contour encloses approximately the searched area. Crosses mark Zwicky clusters, diamonds mark Abell clusters.

tween 180° ≤ l ≤ 185° and between 195° ≤ l ≤ 205°. Around l=210° b >5° there is an empty region with no obvious nebula associated with it. Below the plane more objects were found around l=190°. The region includes 19 Zwicky clusters which have ranges in population sizes from 58 to 403. We did not attempt to measure all objects within the Zwicky clusters but rather some representatives within the general outline of the cluster were measured (the brighter, larger objects). These clusters are marked by crosses in figure 2. We have also plotted as a reference in this figure the centers of Abell clusters (shown as diamonds).

To study the properties of extinction in this area we combined the HI data from the surveys by Heiles(1975) and Weaver & Williams(1973). Figure 3 shows the B-band extinction computed following Burstein & Heiles (1978). The estimated extinction can be up to 4 magnitudes in some regions. At lower latitudes the extinction increases. Around 200° the extinction above the plane tends to be less than below the plane. Around 180° the extinction tends to be similar above and below the plane. The plot serves to show that extinction is very irregular. The distribution of galaxy candidates is clearly correlated with the extinction.

4.   Observations

We obtained 21cm spectra of 373 galaxy candidates from the catalog. This subsample was chosen by considering the available observing time and within that range we selected objects with the better images. In Figure 4 the upper histogram shows the distribution in sizes for ther whole catalog and below is the

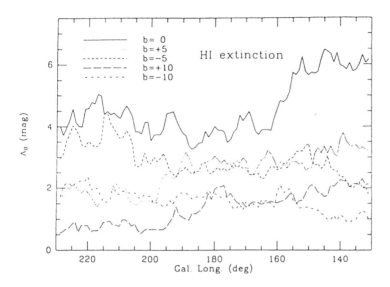

Figure 3.   Extinction in the region studied. The B-band extinction is plotted for intervals of 5° in galactic latitude as a function of longitude.

distribution for the selected sample. The typical size for galaxy candidates in the catalog is 0.3mm. The observed sample also peaks around 0.3mm, while the distribution in sizes for the detected objects peaks at 0.4mm. From this figure it is clear that the fraction of detections increases with image size.

We used the 22cm circular polarization feed, which yields a gain better than 5 K/Jy on two independent circularly polarized components of the incoming radiation, over the frequency range between 1300 and 1415 MHz. The instantaneous bandwidth coverage of the feed-back end configuration was 35 MHz, so that several separate observations were necessary to cover the likely redshift range of each candidate galaxy. The spectral resolution was 8 km s^{-1}, reduced to about $16kms^{-1}$ after hanning smoothing. The spectra were obtained by observing 5 min on-source and 5 min off-source tracking the same antenna pointing. The search was done in such a way that we observed one on-off pair in a low velocity range ($-350kms^{-1} \leq v \leq 7100kms^{-1}$) and if there was no detection then two on-off pairs were made at a high velocity range ($6200kms^{-1} \leq v \leq 13800kms^{-1}$). One object would require on average between 10 to 20 min of integration time. Sometimes due to interference, more scans were necessary for the object. The data was reduced using the Arecibo Observatory standard data reduction package GALPAC.

5.   Results

We detected 137 galaxies from a sample of 373 candidates for a detection rate of $\approx 37\%$. Figure 5 shows the distribution in galactic coordinates of the subsample and the detections.

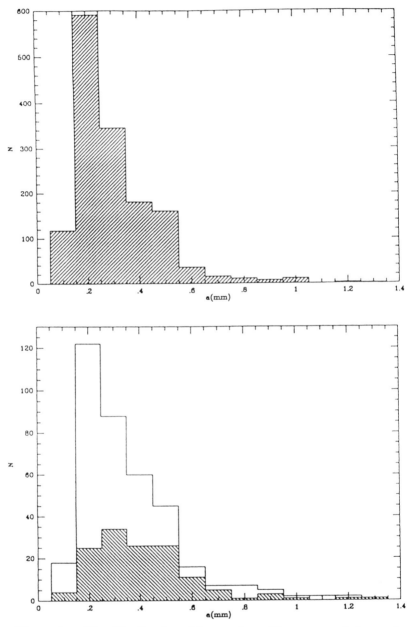

Figure 4. The Distribution of image sizes. The top panel shows the distribution for all objects in the catalog, while the bottom panel shows the same for the observed subsample. Detected objects are shaded on the bottom panel.

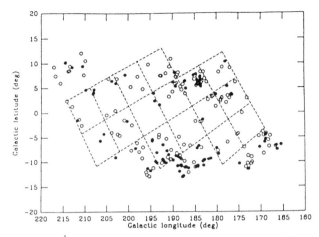

Figure 5.   The observed sample of galaxy candidates. The open circles represent the non-detected objects while the filled circles are the detections. The dashed lines are lines of constant α and δ plotted as a reference.

In addition we observed 173 previously cataloged galaxies (CGCG,UGC,MCG and IRAS PSC) which fall in our region and which did not have a measured redshift. Of these we detected 59. We added these objects and the detections from the Weinberger sample (section 2) to produce the plots shown in figure 6. In these plots we present the heliocentric velocity distribution in right ascension for this augmented sample in the form of cone diagrams for different declination ranges. Figure 6a covers a declination range of $0° \leq \delta < 13°$, figure 6b covers $13° \leq \delta < 26°$ and figure 6c shows $26° \leq \delta < 38°$.

The filled circles are redshifts taken from the CfA redshift catalog (ZCAT, Huchra *et al.* 1992) as published in the ADC CD-ROM, Selected Astronomical Catalogs, Volume 1.The open circles represent the galaxies detected from the sample taken from the optical catalog. The crosses are the detections from the Weinberger sample. The squares represent redshifts from other cataloged galaxies.

A void is evident in figure 6a centered at $\alpha = 8^h.3$ and v $= 6900 km s^{-1}$. Our detections coincide with the outline of this void in the eastern edge. The cluster Abell 539 is located at $\alpha=5^h.3$, $\delta=6°$, v $= 6146 km s^{-1}$ and the cluster Abell 400 is found at $\alpha=02^h.9$, $\delta=5^h.8$, v $= 7189 km s^{-1}$.

In figure 6b, the feature at $\alpha = 7^h$ is constituted mostly of CGCG galaxies with a range in declination from $13° \leq \delta \leq 21°$ and is not a cluster. The cluster AWM 1 appears at $\alpha = 9^h.25$, $\delta = 20°$, v $= 9000 km s^{-1}$, and the Cancer cluster is seen at $\alpha = 8^h.3$, $\delta = 21°$, v $= 5000 km s^{-1}$. We found most of the detected objects in this declination range in the region with $5^h \leq \alpha \leq 6^h$ and $5000 km s^{-1} \leq v \leq 10000 km s^{-1}$.

The cluster Abell 779 is located at $\alpha = 9^h.3$, $\delta = 34°$, v $= 6775 km s^{-1}$ (figure 6a). A density enhancement due to the Pisces-Perseus supercluster is

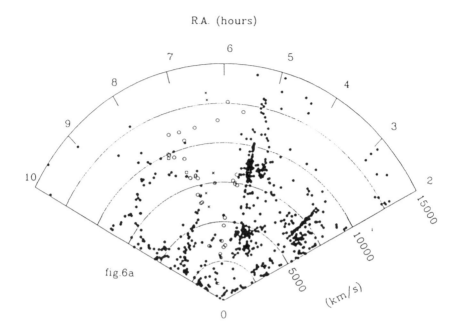

fig.6a

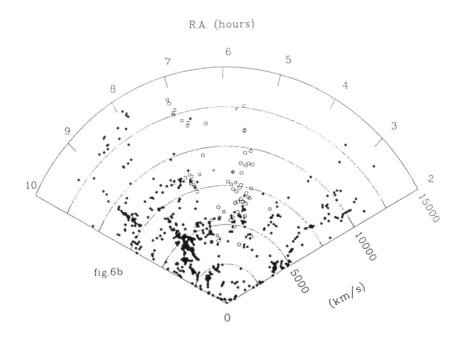

fig.6b

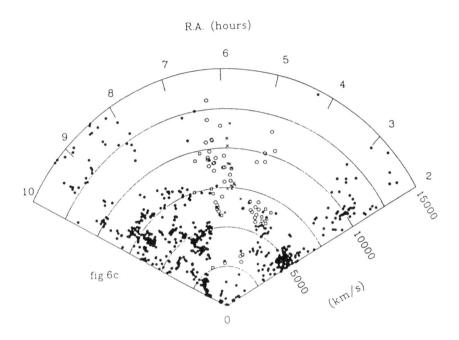

Figure 6.   Distribution of Galaxies in redshift space as a function of right ascension for three different declination ranges.Filled circles are redshifts taken from the ZCAT (Huchra *et al.* 1992), open circles detections from the optical catalog, crosses are Weinberger galaxies, squares are redshifts for other cataloged galaxies. 6a) $0° \leq \delta < 13°$ 6b) $13° \leq \delta < 26°$ 6c) $26° \leq \delta < 38°$.

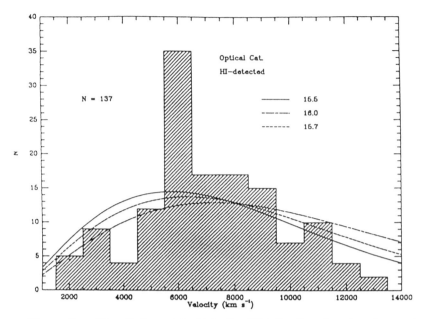

Figure 7.   The Heliocentric Velocity Distribution for the observed
sample from the Optical catalog.

seen appearing at $\alpha = 2^h$, v $= 5000kms^{-1}$ whereas no galaxies were found at
$5000kms^{-1}$, $5^h \leq \alpha \leq 6^h$. At $\alpha = 6^h.25$, v $= 6250kms^{-1}$ there is a group of
galaxies that have a range in declination $26° \leq \delta \leq 29°$.

A region devoid of galaxies, centered at about $\alpha = 6^h$, v $= 10000kms^{-1}$,
appears in panels 6a and 6b. Galaxies were observed in the foreground and in
the background in this direction. The region clearly ends as we get to the third
panel at high declinations.

Figure 7 shows a histogram of the heliocentric velocity distribution.This
plot shows that most of the detected galaxies are found at a redshift of about
$6000kms^{-1}$.

6.   Conclusions

We are currently pursuing the analysis of our data which once completed will
provide 220 new redshifts in this region of the ZOA. Although the search covered
the distance range $0 \leq D \leq 130h^{-1}$ Mpc we found no objects below 1165km s^{-1}
so that our search did not find any new member of the Local Group.

We found the redshifts of our detected objects peaked at about $6000kms^{-1}$
(fig.7). It should be noted that within the region explored lies Abell cluster
539 with $\alpha = 05^h$ 13^m $54^s.0$, $\delta = +06°$ $24'.0$ and v $= 8510kms^{-1}$ (Postman,
Huchra & Geller, 1992). The Great Wall approaches this region of the ZOA
from the west. At the westernmost side we find Abell 779 with coordinates $\alpha =$
09^h 16^m 48^s, $\delta = +33°$ $59'.0$ and at a redshift of $6775kms^{-1}$. At the east but

at higher declinations we have the Pisces-Perseus supercluster approaching this area of the ZOA. The enhancement of galaxies detected around $6000 kms^{-1}$ is compatible with a connection of Pisces-Perseus with Abell 569. This possibilty has been indicated by Chamaraux, *et al.* (1990).

We are also interested in investigating the nature of the non-detected objects which if not galactic might be galaxies out of the velocity range searched or early-type galaxies. It would therefore be interesting to obtain optical redshifts for the non-detections.

Acknowledgments. This research has made use of the NASA/IPAC Extragalactic Database (NED) which is operated by the Jet Propulsion Laboratory, California Institute of Technology, under contract with the National Aeronautics and Space Administration.

C.A.P. acknowledges the NAIC for support that made it possible to participate in this workshop.

References

Beichman, C.A., Neugebauer, G., Habing, H.J., Clegg, P.E., and Chester, T.J. 1988, IRAS Point Source Catalog, NASA RP-1190, Washington D.C.

Brotzman, L.E., and Gessner, S.E., Astronomical Data Center, Selected Astronomical Catalogs, Volume 1

Burstein, D., and Heiles, C. 1978, Ap.J., 225, 40

Chamaraux, P., Cayatte, V., Balkowski, C., and Fontanelli, P. 1990, Astron. Astrophys., 229, 340

Dixon, R.S.,Gearhart, M.R., and Schmidtke, P.C. 1981, Atlas of Sky Overlay Maps,Published by the Ohio State University Radio Observatory

Dow, M.W., Lu, N.Y., Houck, J.R., and Salpeter E.E. 1988, Ap.J., 324, L51

Giovanelli, R., and Haynes, M.P. 1982, Astron.J., 87, 1355

Giovanelli, R., Haynes, M.P., and Chincarini, G.L. 1986, Ap.J., 300, 77

Heiles, C. 1975, Astron.Astrophys.Suppl., 20, 37

Herter, T. 1992, (private communication)

Huchra, J.P., Geller, M.J., Clemens, C.M., Tokarz, S.P., and Michel A. 1992, Harvard-Smithsonian Center for Astrophysics

Pantoja, C.A., Altschuler, D.R., Giovanardi, C., and Giovanelli R. 1994, (in preparation)

Pantoja, C.A., Giovanardi,C., Altschuler, D.R., and Giovanelli, R. 1994, Astron.J., (submitted)

Postman, M., Huchra, J.P., and Geller, M.J. 1992, Ap.J., 384, 404

Weaver, H., and Williams, D.R.W. 1973, Astron.Astrophys.Suppl., 8, 1

Weinberger, R. 1980, Astron.Astrophys.Suppl., 40, 123

Discussion

A. Fairall: Your peak at 7000 km s^{-1} is higher than the usual 5000 km s^{-1}. Perhaps, I could direct a question to you and to John Huchra: Does this suggest that Perseus-Pisces also has something of a 'great wall' structure - that we are seeing edge-on?

C. Pantoja: Although our peak at 7000 km s^{-1} could be related to the splitting of the 'great wall' to the north of the ZOA, we believe that detailed quantitative analysis of our data, when combined with that of other investigations, will lead to more definitive answers about structure in this region.

N. Lu: Comment to the question asked by A. Fairall. The Great Wall appears to diffuse out after it crosses the ZOA to the Pisces-Perseus region.

J. Huchra: One thing I would like to point out with regard to this data and an earlier question is that there is a cluster (Abell 539) in this region at 5^{h}30, +6°, which is at a redshift of about 7000 km/s. Perhaps the redshift peak you see at 7000 is associated with this cluster rather than a high velocity extension of Pisces-Perseus.

C. Balkowski: Do you plan to do optical observations of the objects you did not detect in HI?

C. Pantoja: It would be very interesting to further investigate the nature of the non-detected objects, whether they are early type galaxies or have redshifts outside of the range we searched, and we indeed would like to follow up with optical observations.

N. Lu: Can you talk a little about the IRAS sample you observed in HI?

C. Pantoja: Using the criteria published in your 1990 paper, we selected all those objects from the PSC and made 21-cm observations. We did not have a very high detection rate for these objects, probably because of the high contamination at these low latitudes.

F. Kerr: Can you say a little about the blind search that you are doing?

C. Pantoja: In this form of observing, the telescope is set at a fixed declination and the observations are made as the sky drifts by. The observations are done in total power mode. The ON has an integration time of 10 s and the OFF consists of the mean of 150 spectra. We have completed several strips during different observing sessions and are reducing this data.

Results from Multifiber Spectroscopy of Galaxies in the Zone of Avoidance

V. Cayatte and C. Balkowski

Observatoire de Paris, DAEC, Unité associée au CNRS, DO 173, et à l'Université PARIS VII, 92195 Meudon Cedex, France

R.C. Kraan-Korteweg

Kapteyn Astronomical Institute, Postbus 800, NL-9700 AV Groningen, The Netherlands

Abstract. We have done multi-object spectroscopy with the new instrument MEFOS at ESO to study the galaxy distribution in the Zone of Avoidance.

1. Observations

During 2.5 nights in February 1993 we have observed nine fields located in the Zone of Avoidance close to the Hydra and Antlia clusters (cf. Kraan-Korteweg et al. 1994, this volume) with the new Multifiber Spectroscopy System MEFOS (Felenbok 1992).

MEFOS is mounted at the prime focus of the ESO 3.60m telescope and uses the Boller and Chivens spectrograph. 29 galaxy spectra can be obtained simultaneously in a field of 1° in diameter. Because of arm positioning constraints, not all the objects can always be observed in one go. This can, of course, be resolved through further exposures of the same field on the remaining objects. Before the spectroscopic exposures, images of the objects in the field are made with the image fiber bundle. This allows optimal recentering of the object on the spectroscopic fiber. Typical exposure times were 2x30min. With the grating chosen here we obtained a dispersion of about 170Å/mm and covered the wavelength range from 3950 to 6280Å.

2. Results

In order to give an idea about the instrumental performances we present here the results for 4 typical galaxies from a field located at $\ell = 279°3$ and $b = -8°1$. The image and the spectrum for these 4 galaxies are shown in figure 1. The B_J magnitudes, diameters and the surface-brightness – indicated in Fig. 1 – are estimates from the ESO/SERC IIIaJ film copies made by Kraan-Korteweg (cf. Kraan-Korteweg 1989, or Kraan-Korteweg and Woudt 1994 this volume, for further details). Fig. 1 illustrates quite well that the central surface brightness (not the magnitude) is the dominant factor in achieving a high signal to noise spectrum.

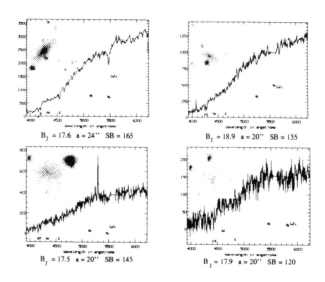

Figure 1.   The 0.'5x0.'5 CCD images and spectra for four typical objects in a MEFOS field. The magnitude (B_J), diameter, and surface-brightness of the galaxies are indicated.

On the 6 fields reduced to date we have observed 113 galaxies. We could determine a radial velocity for 99 of them. Comparing the magnitude, diameter and surface-brightness distributions of the 113 observed galaxies with that of the galaxies for which a radial velocity could be extracted reveals no apparent difference. This shows that we do obtain redshifts for quite faint ($B_J \lesssim 19^m.0$) and low surface-brightness galaxies if they have a relatively bright nucleus. Considering the fact that we are observing fields in the Milky Way, the few confusion problems with foreground stars are amazing: of the 113 spectra, only 9 showed a superimposed star spectrum. And for two of those, it still was possible to deduce a redshift.

3. Conclusion

Based on these data it is evident that the 3.60m ESO telescope equipped with the MEFOS system is an ideal tool for obtaining redshifts of galaxies located in the Milky Way.

References

Felenbok, P., 1992, in Proceedings of the XIIth Moriond Astrophysical Meeting, "Physics of Nearby Galaxies: Nature or Nurture", ed. Thuan et al., p.477

Kraan-Korteweg, R.C. 1989, Rev.Mod.Astr, 2, ed. G. Klare, Berlin Springer, p.119

Discussion

F. Kerr: How long does it take to set up the 29 fibres in their intended positions? Is the setting-up time a significant fraction of the whole observation time?

P. Felenbok: As a first guess, we need 11 min from parking position to the start of a spectrum for B=17 with a spectroscopy exposure of 1 hour. For B=20, it will be 15 min for a spectroscopy exposure of 1h30 min. This is between 15 and 20% of total time.

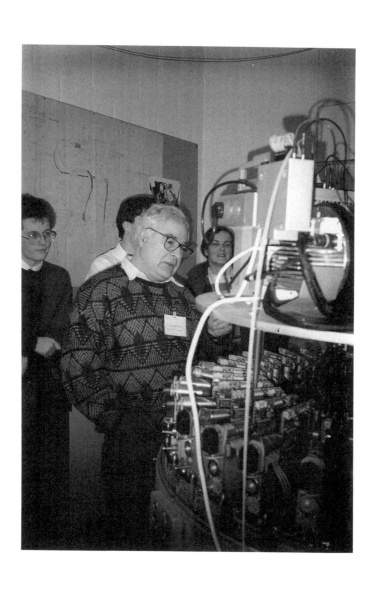

Unveiling Large-Scale Structures Behind the Milky Way
ASP Conference Series, Vol. 67, 1994.
C. Balkowski and R. C. Kraan-Korteweg (eds.)

A Presentation of Galaxies Identified in the Region
$120° \leq \ell \leq 130°$, $-10° \leq b \leq +10°$.

Georg Lercher

Institut für Astronomie der Universität Innsbruck, Technikerstraße 25, 6020 Innsbruck, Austria

Abstract. As a contribution to unveil the extragalactic sky behind the Northern Galactic Plane I searched for optically detectable galaxies on POSS E (red-sensitive) prints in the region of interest. Altogether, 1162 galaxies, the vast majority of them new, could be discovered. The objects were cross-checked with the IRAS-PSC.

1. Introduction

My contribution to the ZoA research is to be seen within the framework of the Innsbruck ZoA project (see this volume for further details), which comprises the entire Northern Galactic Plane. I have chosen an area of $200\square°$, extending from $\ell = 120°$ to $130°$, and $b = -10°$ to $+10°$. The main reasons for my choice were as follows:

1. The region is inbetween the galactic center and the anticenter, where a statistically relevant but not too large galaxy sample can be expected.

2. In this very region, at about $125°$ to $130°$, we can anticipate galaxies even very close to the galactic plane, as was demonstrated by Weinberger (1980, figure 1).

3. The supergalactic plane crosses the galactic plane at about $\ell = 137°$, i.e. near the region I have chosen.

4. The famous Maffei galaxies (Maffei 1 and 2) are located only a few degrees away; it could therefore not be excluded that another nearby object(s) might be detectable.

2. Method

Due to the fact that only few film copies of the POSS-II atlas are available yet , I inspected POSS-I red-sensitive (E) prints by aid of a binocular microscope with a 16-fold magnification and marked all those objects that were thought by me to be of extragalactic nature as well as faint sources of other kinds that might have escaped attention of earlier surveyors.
As a second step, all objects marked by me were checked by another person experienced in optical identifications (Weinberger). A lower limit for the galaxies diameter of 0.1 arcmin was chosen.

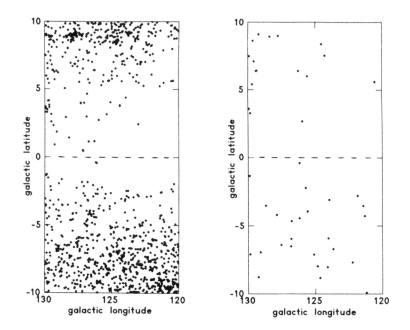

Figure 1.   The distribution of the 1162 optically identified galaxy candidates (left) and the IRAS-PSC identifications (right).

3.   Results of the survey

1. The survey on the POSS-I(E) prints led to the discovery of altogether 1162 galaxy candidates, 159 of them located at $|b| \leq 5°$. Fig.1 shows the distribution of these sources. There is a striking offset of the 'dust equator' compared with the prescribed galactic equator. As to galactic longitude, we note a largely homogeneous distribution, but also some density enhancements at about (ℓ, b) $(130°, 4°)$ and $(128°, 9°)$, where the former might represent a hole in the dust layer or a galaxy cluster, and the latter almost certainly a galaxy cluster.

2. Cross-Identification with the IRAS-PSC:
   43 of the 1162 objects turned out to be IRAS sources. This corresponds to a percentage of about 4%.

Acknowledgments.   This work was partially supported by the Fonds zur Förderung der wissenschaftlichen Forschung, Projekt Nr.P8325-PHY.

References

Weinberger R.: 1980, A&AS **40**, 123

Discussion

D. Altschuler: Very interesting. How do you plan to follow up on this work? Any spectroscopy?

G. Lercher: It would be very interesting to do spectroscopy or 21-cm measurements, as some interesting density enhancements are visible in the surveyed region. As Austria is not member of ESO, we have only few observing possibilities, so we would be very interested in collaboration with other people.

Unveiling Large-Scale Structures Behind the Milky Way
ASP Conference Series, Vol. 67, 1994.
C. Balkowski and R. C. Kraan-Korteweg (eds.)

A search for galaxies behind the galactic plane at l=220

A.I. Terlevich

1 Bradrushe fields, Cambridge, England. CB3 0DW

1. Discussion

The region near the galactic plane at l=220 b=0 was searched for galaxies in two strips, at declinations of d=-5 and d=-10 degrees. This region was chosen as there has been very little work carried out in this area before, owing to its being in the gap between the UGC and ESO catalogues, and also in the zone of avoidance of the Milky Way.

A total of 10 plates from the PSS were visually scanned. Table 1 is a list of the plates scanned. The total area of sky covered was approximately 324 square degrees, and a total of 480 galaxies were found.

The positions of the galaxies were measured with a Coradograph. All galaxies with a major axis larger than 30 arcsec were included in order to have a high confidence of completeness for larger galaxies. The diameters were roughly measured using a ruler photocopied onto a transparency.

As this was a visual survey, some systematic error between plates and even across plates was inevitable. They were minimised by scanning the plates in a random order where possible and using a transparent grid as a guide. The galaxies on the edge of the plates appear on more than one, so their positions were averaged out for the final list.

A raster scan of each galaxy was made in order to determine their radii with some accuracy, however they have not yet been analysed.

A total of 480 galaxies were found on the 10 plates scanned. There is a rapid drop in galaxies found right in the galactic plane as can be seen from the graph where the galactic plane (b=0) is marked, unfortunately, without accurate information on the radii of the galaxies it is not possible to say anything about the distributions, such as the much higher density of galaxies in plates 773 and 774, as they could just be selection effects.

By comparing overlapping plates it can be seen that there are many galaxies in the overlapping regions that don't appear on both plates when scanned indicating that the scan isn't complete, however the galaxies that were missed out were practically all on the limit of being included in the searc due to their having radii of about 30 arcsec. None of these missed out galaxies had a radius larger than 1 arcmin.

Acknowledgments. This work was done while I was a student at the RGO/IOA summer school, under the supervision of Donald Lynden-Bell, Ofer Lahav and Caleb Scharf.

Table 1.   Plates searched showing cental position of each and the number of galaxies found.

Plate	RA	DEC	Number of galaxies
PSS 770	$6^h\ 20^m$	-5°	38
PSS 696	$5^h\ 40^m$	-10°	60
PSS 700	$7^h\ 00^m$	-10°	0
PSS 699	$6^h\ 40^m$	-10°	17
PSS 698	$6^h\ 20^m$	-10°	42
PSS 697	$6^h\ 00^m$	-10°	61
PSS 771	$6^h\ 40^m$	-5°	11
PSS 772	$7^h\ 00^m$	-5°	3
PSS 773	$7^h\ 20^m$	-5°	162
PSS 774	$7^h\ 40^m$	-5°	128

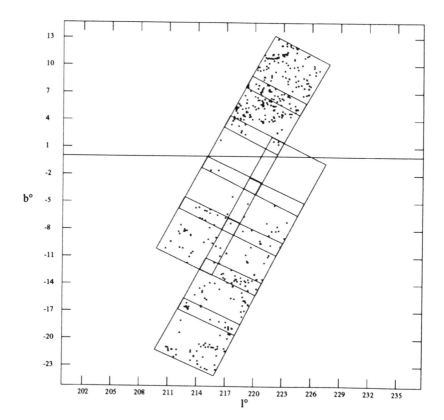

Figure 1.   Graph showing the positions of the plates searched and the galaxies found in galactic coordinates.

Discussion

N. Lu: What fraction of the contrast in galaxy number between above and below the plane is due to galactic extinction?

D. Lynden-Bell (Terlevich): Certainly the 21-cm data give a steep fall off in the north where the galaxies are thickest and a slow fall off in the south. I do not know whether there is an enthusiasm effect on how deeply the plates were searched.

Unveiling Large-Scale Structures Behind the Milky Way
ASP Conference Series, Vol. 67, 1994.
C. Balkowski and R. C. Kraan-Korteweg (eds.)

STUDY OF THE SHAPLEY CONCENTRATION

SANDRO BARDELLI
Dipartimento di Astronomia, via Zamboni 33, 40126 Bologna (Italy)
ELENA ZUCCA
Istituto di Radioastronomia, CNR, via Gobetti 101, 40129 Bologna (Italy)
ROBERTO SCARAMELLA
Osservatorio Astronomico di Roma, 00040 Monteporzio Catone (Italy)
GIAMPAOLO VETTOLANI
Istituto di Radioastronomia, CNR, via Gobetti 101, 40129 Bologna (Italy)
GIOVANNI ZAMORANI
Osservatorio Astronomico, via Zamboni 33, 40126 Bologna (Italy)
Istituto di Radioastronomia, CNR, via Gobetti 101, 40129 Bologna (Italy)

ABSTRACT   We present first results of an extensive study of the Shapley Concentration aimed to describe its dynamical state and to determine its total mass.

The Shapley Concentration is the most remarkable feature which appears studying the distribution of the Abell–ACO clusters of galaxies (Scaramella et al. 1989). It has also a great relevance in the peculiar motion problem: indeed it may be responsible for $\sim 30\%$ of the acceleration acting on the Local Group of galaxies (Scaramella et al. 1991). This supercluster stands out also studying the bi–dimensional distribution of optical galaxies (Raychaudhury et al. 1991), analysing the spatial distribution of IRAS galaxies (Allen et al. 1990), and it is also prominent in the X–ray band (Lahav et al. 1989). The central part of this supercluster is formed by two condensations: $A3556 - A3558 - A3562$ and $A3528 - A3530 - A3532$. In particular, the three clusters $A3556 - A3558 - A3562$ form a structure elongated for $\sim 3^o$ along the East–West direction and can be considered as the core of the Shapley Concentration (see Fig. 1).

We are carrying on an extensive study of the Shapley Concentration in order to describe its dynamical state and to determine its total mass. The core of this supercluster was observed in several runs with the multifiber spectrograph OPTOPUS at La Silla: we have obtained a sample of ~ 500 new galaxy redshifts. These data, added to the redshifts already present in literature, lead to a total sample of 681 galaxies with measured velocity (Bardelli et al. 1994). We found that the clusters in this chain are each other interacting and form a single, connected structure; in particular, $A3558$ presents many subcondensations and could be accreting galaxies from nearby groups and clusters, as confirmed also by a pointed ROSAT PSPC observation (Bardelli et al., in preparation): probably this is the beginning of a merging process. We are now reducing ~ 500 redshifts in the $A3528$ complex, which is a structure with similar properties.

Finally, we are studying the whole supercluster, considering in particular the inter–cluster galaxies. Indeed from the number–magnitude counts on 12 UKSTJ plates, derived from the COSMOS scans, we have found the presence of a galaxy excess in the Shapley Concentration, even if we eliminate galaxies

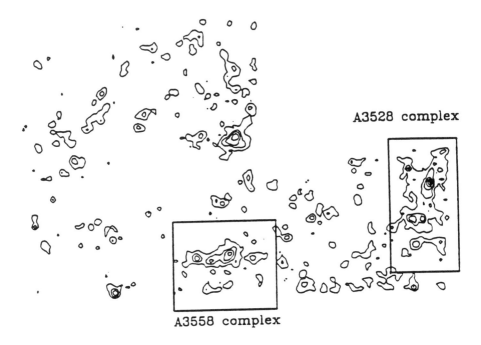

A3528 complex

A3558 complex

Fig.1: Isodensity contours of the galaxies in central region of the Shapley Concentration. The isodensity levels have been chosen in order to better visualize the clusters.

belonging to clusters. Therefore we are now measuring redshifts, with the multifiber spectrograph MEFOS at La Silla, of "field" galaxies in order to derive the radial profile of the whole supercluster. These information will allow us to estimate the extension and the overdensity in galaxies of the Shapley Concentration, and therefore its real contribution to the local velocity field.

REFERENCES

Allen, D.A., Norris, R.P., Staveley–Smith, L., Meadows, V.S., and Roche, P.F., 1990, *Nature* **343**, 45.

Bardelli et al., 1994, *M.N.R.A.S.*, in press.

Lahav, O., Edge, A.C., Fabian, A.C., and Putney, A., 1989, *M.N.R.A.S.*, **238**, 881.

Raychaudhury, S., Fabian, A.C., Edge, A.C., Jones, C., and Forman, W., 1991, *M.N.R.A.S.*, **248**, 101.

Scaramella, R., Baiesi–Pillastrini, G., Chincarini, G., Vettolani, G., and Zamorani, G., 1989, *Nature*, **338**, 562.

Scaramella, R., Vettolani, G., and Zamorani, G., 1991, *Ap. J. (Letters)*, **376** 1.

Part IV

Theoretical Presentations

Unveiling Large-Scale Structures Behind the Milky Way
ASP Conference Series, Vol. 67, 1994.
C. Balkowski and R. C. Kraan-Korteweg (eds.)

Wiener Reconstruction of All-Sky Spherical Harmonic Maps of the Large Scale Structure

Ofer Lahav

Institute of Astronomy, Madingley Road, Cambridge CB3 0HA, UK

Abstract. A statistical method for reconstructing large scale structure behind the Zone of Avoidance is presented. It also corrects for shot-noise and for redshift distortion in galaxy surveys. The galaxy distribution is expanded in an orthogonal set of spherical harmonics. We show that in the framework of Bayesian statistics and Gaussian random fields the 4π harmonics can be recovered and the shot-noise can be suppressed, giving the optimal picture of the underlying density field. The correction factor from observed to reconstructed harmonics turns out to be the well-known Wiener filter (the ratio of signal to signal+noise), which is also derived by requiring minimum variance. We apply the method to the 1.2 Jy IRAS survey. A reconstruction of the projected galaxy distribution confirms the connectivity of the Supergalactic Plane across the Galactic Plane (at Galactic longitude $l \sim 135°$ and $l \sim 315°$) and the Puppis cluster behind the Galactic Plane ($l \sim 240°$). The method is extended to 3-D, and is used to recover from the 1.2 Jy redshift survey the density, velocity and potential fields in the local universe.

1. Introduction

Where the Zone of Avoidance (ZOA) cannot be observed directly, the alternative is to reconstruct the structure in a statistical way. Previous corrections for unobserved regions in catalogues were done, somewhat ad-hoc, by populating the ZOA uniformly according to the mean density, or by interpolating the structure below and above the Galactic Plane (e.g. Lynden-Bell et al. 1989, Yahil et al. 1991, Strauss et al. 1992, Hudson 1993).

Two other problems appear in analysing the distribution of galaxies. First, if one assumes that the distribution of luminous galaxies samples an underlying smooth density field, then the discreteness of objects introduces Poisson 'shot-noise'. Second, due to peculiar velocities, redshift surveys give a distorted picture of the density field. Here we show how to recover all-sky projected density field, from galaxy surveys which suffer incomplete sky, and also how to reconstruct the density, velocity and potential fields from redshift surveys. The results presented here are based on several recent studies, by Lahav et al. (1994), Fisher et al. (1994b) and Zaroubi et al. (1994). Hoffman (this volume) discusses other aspects of the method, mainly when applied in Cartesian coordinates, and Bunn et al. (1994) recently applied a similar method to reconstruct the COBE DMR map.

The recovery of a signal from noisy and incomplete data is a classic problem of inversion, common in problems of image processing (e.g. 'seeing' or blurred HST images). A straightforward inversion is often unstable, and a regularization scheme of some sort is essential in order to interpolate where data are missing or noisy. By avoiding any prior assumptions one allows the noise and incomplete data to sometime dominate the resulting reconstruction. In the Bayesian spirit we use here raw data and a prior model to produce 'improved data'. This may raise the question to what extent a reconstruction of say the Great Attractor depends on what is assumed about the unknown nature of the spectrum of fluctuations. But the prior model does not necessarily require a speculative assumption. In the context of this work we simply require a reconstruction which obeys the constraint of the 2-point correlation function of the observed galaxy distribution, as derived from a smaller section of the sky. Using the above principles we derive a Wiener filter (the ratio of signal to signal+noise), which also follows from requiring minimum variance (e.g. Rybicki & Press 1992).

2. Wiener filter in theory and in practice

Let us first consider a simple pedagogical example. Assume two Gaussian variables, x and y, with zero mean, $\langle x \rangle = \langle y \rangle = 0$ (hereafter $\langle ... \rangle$ denote ensemble average). The probability for x given y is by the rule of conditional probability

$$P(x|y) = \frac{P(x,y)}{P(y)}, \tag{1}$$

where for Gaussian probability $P(y) \propto \exp\left(-\frac{y^2}{2\langle y^2 \rangle}\right)$, and the joint probability is a bivariate Gaussian $P(x,y) \propto \exp\left[-\frac{1}{2}(u^2 - 2\rho uv + v^2)/(1 - \rho^2)\right]$, where $u = x/\langle x^2 \rangle^{1/2}$, $v = y/\langle y^2 \rangle^{1/2}$ and $\rho = \langle xy \rangle/\sqrt{\langle x^2 \rangle \langle y^2 \rangle}$. It then follows that the conditional probability is simply a 'shifted Gaussian'

$$P(x|y) \propto \exp\left[-\frac{1}{2}(u - \rho v)^2/(1 - \rho^2)\right]. \tag{2}$$

The maximum *a poteriori* probability clearly occurs for $\hat{u} = \rho v$, or $\hat{x} = \frac{\langle xy \rangle}{\langle y^2 \rangle} y$. In the special case of Gaussian fields the most probable reconstruction is also the mean field (cf. Hoffman & Ribak 1991; Kaiser & Stebbins 1991). Hereafter we term them together as the 'optimal reconstruction'.

Exactly the same result for the 'optimal reconstruction' is also obtained by a different approach, by asking for the linear filter F which minimizes the variance $\langle (x - Fy)^2 \rangle$. Minimizing with respect to F gives indeed $\hat{F} = \frac{\langle xy \rangle}{\langle y^2 \rangle}$ and $\hat{x} = \hat{F}y$, as above. Note that although the results of the two approaches are identical, due to the quadratic nature of the functions and the linearity of the filter, the underlying assumptions are quite different. The conditional probability approach (eq. 1) requires to specify the full distribution functions (Gaussians in our case). On the other hand, the minimum variance approach only considers the second moment of the distribution function, but assumes a linear filter F.

Consider now the special case that $y = x + \sigma$, where σ is a Gaussian noise uncorrelated with the true signal x (hence $\langle x\sigma \rangle = 0$). It follows that the optimal estimator of the signal $\hat{x}$ given the (noisy) measurement y is

$$\hat{x} = \frac{\langle x^2 \rangle}{\langle x^2 \rangle + \langle \sigma^2 \rangle} \, y \qquad (3)$$

The factor (F) in front of the measurement y is the well-known Wiener filter commonly used in signal processing (Wiener 1949; for review see e.g. Press et al. 1992, Rybicki & Press 1992). Note that it requires *a priori* knowledge of the variances in the signal and the noise. When the noise is negligible the factor approaches unity, but when it is significant the measurement is attenuated.

A third approach is of adding a regularizing function to the usual χ^2 (log-Likelihood) minimization. In fact a regularization function of the form x^2, motivated in our case by physical considerations of the underlying field, yields essentially a Wiener filter. Other reconstruction method use different regularization functions, e.g Maximum Entropy (e.g. Gull 1989) takes $x \ln x$.

Here we have only considered a simple example of two variables. More generally, for vectors of signal and noise and a response function $W_{\alpha\beta}$ (e.g. a 'point spread function') one can write $y_\alpha = W_{\alpha\beta}[x_\beta + \sigma_\beta]$ and derive a Wiener solution of the form $\hat{x}_\alpha = \langle x_\alpha y_\gamma \rangle \langle y_\gamma y_\beta^\dagger \rangle^{-1} y_\beta$. The Wiener formulation is greatly simplified by using orthogonal set of functions, e.g. by employing a Fourier or harmonic transforms. In particular, if $W_{\alpha\beta} = 1$ then the Wiener matrix in Fourier space is diagonal, and eq. (3) holds, but with the variables replaced by their Fourier transforms.

Figure 1 shows a 1-d example, which is also of relevance to the ZOA problem. The solid line at the bottom panel is a mock 'double-horned' HI spectrum of a galaxy (generated by H. Ferguson). To this we added real noise taken from the Dwingeloo radio-telescope, resulting in a noisy spectrum at the top panel. I then applied a Wiener filter in Fourier space, using the prior rms of the galaxy and the noise (here of course we know what they are). The dotted line in the bottom panel shows the Wiener reconstruction which indeed recovers reasonably well the galaxy spectrum. In fact, we are developing this Wiener approach as a detection algorithm for the Dwingeloo project of HI blind search behind the ZOA.

3. Expansion in Spherical Harmonics

Back to the large scale structure, in analysing galaxy surveys the most informative data set is of course the catalogue itself. However, it is more efficient and sometimes more insightful to compress the galaxy data. Here we use spherical harmonics to expand the galaxy distribution in a whole-sky survey. This technique has been considered for 2-D samples (e.g. Peebles 1973, Scharf et al. 1992) and more recently for analysing redshift and peculiar velocity surveys (e.g. Regös & Szalay 1989; Scharf & Lahav 1993; Lahav et al. 1993; Fisher, Scharf & Lahav 1994a; Nusser & Davis 1994; Lahav 1994 for a summary of properties).

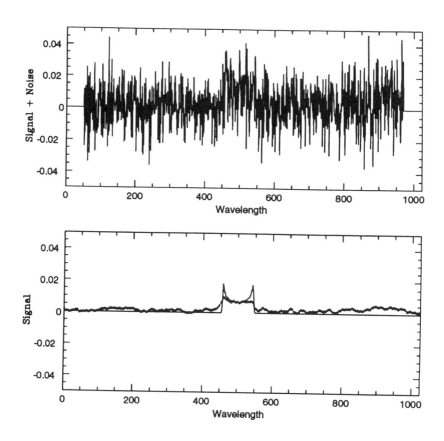

Figure 1.   HI spectrum of a mock galaxy+noise (top panel) and its Wiener reconstruction (dotted line, bottom panel), compared with the original mock galaxy spectrum (solid line, bottom line). The units are arbitrary.

In projection, the density field over 4π is expanded as a sum:

$$S(\theta, \phi) = \sum_l \sum_{m=-l}^{m=+l} a_{lm} \, Y_{lm}(\theta, \phi), \tag{4}$$

where the Y_{lm}'s are the orthonormal set of spherical harmonics : $Y_{lm}(\theta, \phi) \propto P_l^{|m|}(\cos\theta) \exp(im\phi)$, where θ and ϕ are the spherical polar angles, and $P_l^{|m|}$'s are the associated Legendre Polynomials of degree l and order m. The spherical harmonic analysis provides a unified language to describe the local cosmography as well as the statistical properties (e.g. the power-spectrum) of the galaxy distribution. In particular it retains both the amplitude and phase information and hence the underlying texture of the distribution.

The mean-square of harmonics can be related to the power-spectrum of mass fluctuations $P_m(k) = \langle |\delta_k|^2 \rangle$ in Fourier space. In particular, the variance in harmonics measured in a flux limited redshift survey can be formulated (Fisher, Scharf & Lahav 1994a) assuming linear theory as

$$\langle |a_{lm}^S|^2 \rangle = \frac{2}{\pi} \, b^2 \int dk \, k^2 P_m(k) \left| \Psi_l^R(k) + \frac{\Omega_0^{0.6}}{b} \Psi_l^C(k) \right|^2, \tag{5}$$

where $\Psi_l^R(k)$ and $\Psi_l^C(k)$ are 'window functions'. By applying this relation to the 1.2 Jy redshift survey (of Fisher 1992, Strauss et al. 1992) we find for the combination of density and bias parameters, $\Omega_0^{0.6}/b \sim 1.0 \pm 0.3$ (assuming the observed IRAS galaxy power-spectrum).

4. Mask inversion using Wiener filter

We turn now to the problem of reconstructing large scale structure behind the ZOA. The problem is formulated as follows: What are the full-sky noise-free harmonics given the observed harmonics, the mask describing the unobserved region, and a prior model for the power-spectrum of fluctuations ?

The observed harmonics $c_{lm,obs}$ (with the masked regions filled in uniformly according to the mean) are related to the underlying 'true' whole-sky harmonics a_{lm} by (cf. Peebles 1980, eq. 46.33)

$$c_{lm,obs} = \sum_{l'} \sum_{m'} W_{ll'}^{mm'} [a_{l'm'} + \sigma_a] \tag{6}$$

where the monopole term ($l' = 0$) is excluded. We have added the shot-noise σ_a in the 'true' number-weighted harmonics a_{lm}'s (not in the c_{lm}'s). The noise variance is estimated as $\langle \sigma_a^2 \rangle = \mathcal{N}$ (the mean number of galaxies per steradian, independent l in this case). The harmonic transform of the mask, $W_{ll'}^{mm'}$, introduces 'cross-talk' between the different harmonics.

By analysis similar to that given in section 2 it can be shown (Lahav et al. 1994; Zaroubi et al. 1994) that the solution of this inversion problem is

$$\hat{\mathbf{a}} = \mathbf{F} \mathbf{W}^{-1} \mathbf{c}_{obs}, \tag{7}$$

where the vectors **a** and c_{obs} represent the sets of observed harmonics $\{a_{lm}\}$ and $\{c_{lm,obs}\}$, with the diagonal Wiener matrix

$$F = diag\Big\{ \frac{\langle a_l^2 \rangle_{th}}{\langle a_l^2 \rangle_{th} + \langle \sigma_a^2 \rangle} \Big\}. \tag{8}$$

Here $\langle a_l^2 \rangle_{th}$ is the cosmic variance in the harmonics, which depends on the power-spectrum (cf. eq. 5). We emphasize again that in the special case of underlying Gaussian field the most probable field, the mean field and the minimum variance Wiener filter are all identical. The scatter in the reconstruction is at least as important, and it can also be written analytically for Gaussian random fields.

Even if the sky coverage is 4π (**W** = **I**), the Wiener filter is essential to reveal the optimal underlying 'continuous' density field, cleaned of noise. In the absence of other prior information on the location of clusters and voids, the correction factor is 'isotropic' per l, i.e. independent of m, so in the case of full sky coverage, only the amplitudes are affected by the correction, but not the relative phases. For example, the dipole direction is not affected by the shot-noise, only its amplitude. But of course, if the sky coverage is incomplete, both the amplitudes and the phases are corrected. The reconstruction also depends on number of observed and desired harmonics. Note also that the method is *non*-iterative. Since the Wiener factor is less than unity, applying it iteratively will result in zero signal !

5. Reconstruction of the projected IRAS 1.2 Jy galaxy distribution

Here we apply the method to the sample of IRAS galaxies brighter than 1.2 Jy which includes 5313 galaxies, and covers 88 % of the sky. This incomplete sky coverage is mainly due to the Zone of Avoidance, which we model as a 'sharp mask' at Galactic latitude $|b| < 5°$. The mean number of galaxies is $\mathcal{N} = 392$ per steradian, which sets the shot-noise, $\langle \sigma_a^2 \rangle$. As our model for the cosmic scatter $\langle a_l^2 \rangle_{th}$ we adopt a fit to the observed power spectrum of IRAS galaxies (Fisher et al. 1993).

Figure 2(a) shows the reconstruction of the projected IRAS 1.2 Jy sample. The Zone of Avoidance was left empty, and clearly it 'breaks' the possible chain of the Supergalactic Plane and other structures. Figure 2(b) shows our optimal reconstruction for $1 \leq l \leq 15$. Now the structure is seen to be connected across the Zone of Avoidance, in particular in the regions of Centaurus/Great Attractor ($l \sim 315°$), Hydra ($l \sim 275°$) and Perseus-Pisces ($l \sim 315°$), confirming the connectivity of the Supergalactic Plane. We also see the Puppis cluster ($l \sim 240°$) recovered behind the Galactic Plane. This cluster has been noticed in earlier harmonic expansion (Scharf et al. 1992) and other studies (Lahav et al. 1993 and references therein; Yamada in this volume). The other important feature of our reconstruction is the removal of shot noise all over the sky. This is particularly important for judging the reality of clusters and voids.

Comparison of our reconstruction with the one applied (using a 4π Wiener filter) to the IRAS sample in which the ZOA was filled in 'by hand' across the Galactic Plane (Yahil et al. 1991) shows good agreement. We have also used other prior realistic models and found that the reconstructions changed very

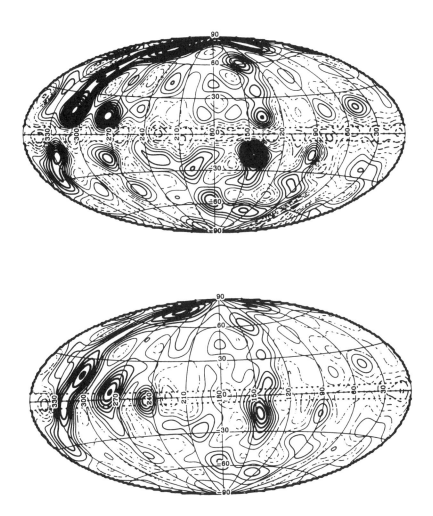

Figure 2. (a) Harmonic expansion ($1 \le l \le 15$) of the projected raw IRAS 1.2 Jy data in Galactic Aitoff projection. Regions not observed, in particular $|b| < 5$ (marked by dashed lines), were left empty. The contour levels of the projected surface number density are in steps of 100 galaxies per steradian (the mean projected density is $\mathcal{N} \sim 400$ galaxies per steradian). (b) A whole-sky Wiener reconstruction of figure (a). The reconstruction corrects for incomplete sky coverage, as well as for the shot-noise. The reconstruction indicates that the Supergalactic Plane is connected across the Galactic Plane at Galactic longitude $l \sim 135°$ and $l \sim 315°$. The Puppis cluster stands out at the Galactic Plane at $l \sim 240°$. The horizontal dashed lines at $b = \pm 5°$ mark the major Zone of Avoidance in the IRAS sample. The contour levels are as in (a). (From Lahav et al. 1994.)

little. As a more challenging test of the method we have also used an N-body simulation of standard Cold Dark matter (where the whole 'sky' true harmonics are known) and varied the size of the ZOA. We find that for mask larger than $|b| = 15°$ it is difficult to recover the unobserved structure. In this case extra-regularization is required, e.g. by truncating components in the Singular Value Decomposition (Press et al. 1992; Lahav et al. 1994). Clearly the success of the method depends on the interplay of three angular scales: the width of the mask, the desired resolution (π/l_{max}) and the physical correlation of structure.

6.   3-D reconstruction of density, velocity and potential fields

To extend the reconstruction method for analyzing redshift surveys we expand the fluctuations in the density field in spherical harmonics Y_{lm} and spherical Bessel functions $j_l(z)$ (cf. Binney & Quinn 1991, Lahav 1994, Fisher et al. 1994b) :

$$\rho(\mathbf{r}) = \sum_l \sum_m \rho_{lm}(r) Y_{lm}(\hat{\mathbf{r}}) = \sum_l \sum_m \sum_n C_{nl}\, \rho_{lmn}\, j_l(k_n r)\, Y_{lm}(\hat{\mathbf{r}})\ , \qquad (9)$$

The harmonics and Bessel functions are natural for this problem as they are the eigen-functions of Poisson equation, and provide a convenient framework for dynamical calculations. The C_{nl}'s define the normalization.

We shall assume that the data are given within a sphere of radius R, such that inside the sphere the desired density fluctuation is specified by $\rho_{lm}(r)$, but for $r > R$ the fluctuation is $\rho_{lm}(r) = 0$ (this simply reflects our ignorance about the density field out there; the fluctuations do not vanish of course at large distances). The Fourier k_n's are chosen to ensure orthogonality of the Bessel functions, by imposing as boundary condition that the logarithmic derivative of the potential is continuous at $r = R$.

An estimator for the density field from the redshift survey is

$$\hat{\rho}^S_{lmn} = \sum_{gal} \frac{1}{\phi(s)}\, j_l(k_n s)\, Y^*_{lm}(\hat{\mathbf{r}}), \qquad (10)$$

where the sum is over galaxies with $r < R$, and $\phi(r)$ is the radial selection function.

Assuming 4π coverage, two corrections are needed in order to convert the redshift space coefficients to noise-free coefficients in real space ρ^R_{lmn}. It is shown in detail in Fisher et al. (1994b) that this can be done by first correcting the density coefficients in redshift space for the distortion (assuming linear theory and $\Omega^{0.6}/b$; cf. eq. 5), and then applying a Wiener filter to remove the shot-noise, assuming a prior for the power-spectrum. Armed with the density coefficients one can then predict (using linear theory) the peculiar velocity field due to the mass distribution represented by galaxies inside the spherical volume. The method provides a non-parametric description of the density, velocity and potential fields, which are related by simple linear transformations. Figure 3 shows a reconstruction by this method using the IRAS 1.2 Jy redshift survey, a prior IRAS power-spectrum and $\Omega^{0.6}/b = 1$. Here we have used the IRAS survey with the ZOA filled in by the interpolation of Yahil et al. (1991), which gave

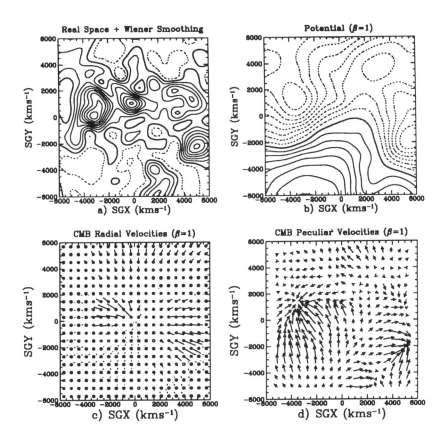

Figure 3.   (a) The reconstructed real space density field from the 1.2 Jy IRAS redshift survey in the Supergalactic Plane. Contours are spaced at $\Delta\delta = 0.5$ with solid (dashed) lines denoting positive (negative) contours. The heavy solid contour corresponds to $\delta = 0$. (b) Reconstructed dimensionless gravitational potential field, $\phi(\mathbf{r})/c^2$ from the 1.2 Jy survey for $\beta = 1$. Contours are spaced at $\Delta\phi/c^2 = 5 \times 10^{-8}$. Solid (dashed) contours denote positive (negative) values with the heavy contour representing $\phi = 0$. (c) Reconstructed radial velocity field. Closed (open) dots represent positive (negative) velocities. (d) Reconstructed three dimensional peculiar velocity field. (From Fisher et al. 1994b.)

similar results to our Wiener projected reconstruction described above. While the 3-D method can be extended to account for the incomplete sky coverage, it was more convenient mathematically in this case (with relatively small of ZOA and harmonics $l < 15$) to use the interpolated data and to formulate the problem for 4π. It is remarkable that the two major dynamical features in the map, the Great Attractor region (at $SGX \sim -3500$ km/sec; $SGY \sim 0$ km/sec in the density map) and the Perseus-Pisces supercluster (at $SGX \sim 5000$ km/sec; $SGY \sim -2000$ km/sec) are both very near the ZOA.

7. Discussion

We have presented Wiener filtering method for reconstructing the full sky density, velocity and potential fields, free of shot-noise. We have also shown that a variety of statistical approaches to the problem all lead to the same optimal Wiener estimator. The prior assumptions only depend on the observed 2-point galaxy correlation function and the nature of the shot-noise. Our assumption that the density field is Gaussian and is sampled by luminous galaxies is of course only an approximation to the real universe, but it provides a convenient framework which can be further extended. This method is to be applied to new all-sky IRAS and optical redshift surveys, and to surveys of the peculiar velocity field. The ρ_{lmn}'s coefficients allow objective (non-parametric) comparison of different surveys of light and mass in the local universe. As illustrated here this reconstruction technique can help in probing the ZOA, and in answering some other cosmographic questions.

Acknowledgments. I thank K. Fisher, Y. Hoffman, D. Lynden-Bell, C. Scharf and S. Zaroubi for their contribution to the work presented here and for many stimulating discussions.

References

Binney, J., & Quinn, T. 1991, MNRAS, 249, 678

Bunn, E., Fisher, K.B., Hoffman, Y., Lahav, O., Silk, J. & Zaroubi, S. 1994, preprint

Fisher, K.B. 1992, *PhD thesis*, University of California, Berkeley

Fisher, K.B., Davis, M., Strauss, M. A., Yahil, A., & Huchra, J. P. 1993, ApJ, 402, 42

Fisher, K.B., Scharf, C.A. & Lahav, O. 1994a, MNRAS, 266, 219

Fisher, K.B., Lahav, O., Hoffman, Y., Lynden-Bell, D. & Zaroubi, S. 1994b MNRAS, submitted.

Gull, S.F., 1989, in *Maximum Entropy and Bayesian Methods*, ed. J. Skilling, pg. 53 (Kluwer)

Hoffman Y. & Ribak, E. 1991, ApJ, 380, L5

Hudson, M. 1993, MNRAS, 265, 72

Kaiser, N. & Stebbins, A. 1991, in *Large Scale Structures and Peculiar Motions in the Universe*, pg. 111, eds. D.W. Latham & N. daCosta, ASP Conference Series, vol. 15, (San Francisco)

Wiener Reconstruction

181

Lahav, O. 1994, in *Cosmic Velocity Fields* (Paris, July 1993), eds. F.R. Bouchet
& M. Lachiéize-Rey, (Gif-sur-Tvette Cedex: Editions Frontieres), p. 205

Lahav, O., Yamada, T., Scharf, C., A. & Kraan-Korteweg, R.C. 1993, MNRAS,
262, 711

Lahav, O., Fisher, K.B., Hoffman, Y., Scharf, C.A. & Zaroubi, S. 1994, ApJ,
423, L93

Lynden-Bell, D., Lahav, O. & Burstein, D. 1989, MNRAS, 241, 325.

Nusser, A. & Davis, M. 1994, ApJ, 421, L1

Peebles, P.J.E. 1973, ApJ, 185, 413

Peebles, P.J.E. 1980, *The Large Scale Structure of the Universe*, (Princeton
University Press)

Press, W. H., Teukolsky, S.A., Vetterling, W.T., & Flannery, B.P., 1992, *Numerical Recipes*, 2nd edition, (Cambridge University Press)

Scharf, C.A., Hoffman, Y., Lahav, O., & Lynden-Bell, D. 1992, MNRAS, 256,
229

Scharf, C.A., & Lahav, O. 1993, MNRAS, 264, 439

Strauss, M.A., Yahil, A., Davis, M., Huchra, J.P., & Fisher, K. 1992, ApJ, 397,
395.

Regős, E. & Szalay, A.S. 1989. ApJ, 345, 627

Rybicki, G.B., & Press, W.H. 1992, ApJ, 398, 169

Wiener, N. 1949, *Extrapolation and Smoothing of Stationary Time Series*, (New
York: Wiley)

Yahil, A., Strauss, M.A., Davis, M. & Huchra, J.P., 1991, ApJ, 372, 380

Zaroubi, S., Hoffman, Y., Fisher, K.B & Lahav, O. 1994, in preparation

Discussion

G. Mamon: Can you apply spherical harmonics to one hemisphere of data?

O. Lahav: This is not that practical. The Wiener reconstruction can only recover structure if the missing zone is relatively small and comparable to the correlation scale of the projected galaxy distribution. However, the nice aspect of regularization methods such as Wiener or Maximum Entropy is that they can tell you 'honestly' if the data are bad or missing, and in this case an area like the missing hemisphere is kept 'grey', i.e. at the level of the mean density.

R. Kraan-Korteweg: If the gap due to lack of data in whole-sky samples is too large, your reconstruction method fails. What kind of coverage of the ZOA is required to get a reliable reconstruction of the ZOA? What kind of coverage is needed when you expand your method to 3-D and does the Wiener reconstruction allow the detection of more detailed structure compared to POTENT or Hoffman's method which have very large smoothing (1200 km/s)?

O. Lahav: Our experiments with simulations, projected in the same way as the IRAS survey, indicate that reliable reconstructions are achieved only if $|b| < 15°$. In 3-D the resolution depends on the number of Bessel radial modes, and this choice depends on the selection function, the shot-noise and the scale of non-linear redshift distortion. Practically, for current surveys 'optimal' smoothing is roughly 500 km/s (Gaussian half-width) at a distance of 4000 km/s, but it increases with distance.

M. Hendry: You find that the minimum variance and maximum probability solutions are equivalent in the Gaussian case. Does this result hold for any symmetric probability distribution? Or, perhaps a more interesting question, does the result fail for any non-symmetric distribution?

O. Lahav: Indeed the Gaussian case is very special in the sense that the mean field is the same as the maximum *a posteriori* solution, and is also the same as the minimum variance solution. This equivalence breaks down for most other distribution functions, e.g. for the log-normal function.

W. Saunders: The Wiener filter leads to unbiased estimates of all the statistical quantities of cosmological interest - variance, velocity distortions etc. However, on a point to point basis the filtered version is always biased towards the mean by an amount dependent on the noise - surely this limits its usefulness for cosmographic purposes.

O. Lahav: First, the Wiener solution, when derived by the minimum variance approach, only guarantees this condition. It may, for example, bias the mean (this can be cured by subtracting the mean before the reconstruction, and adding it back at the end) and high moments. It is true that our formalism assumes that we know the noise properties in the rms, not locally. However, the formalism can be extended to accommodate other prescriptions.

Unveiling Large-Scale Structures Behind the Milky Way
ASP Conference Series, Vol. 67, 1994.
C. Balkowski and R. C. Kraan-Korteweg (eds.)

Wiener Reconstruction of the Large Scale Structure in the Zone of Avoidance

Yehuda Hoffman

Racah Institute of Physics, Hebrew University, Jerusalem, Israel

Abstract. The Wiener filter algorithm is applied to the IRAS $1.9Jy$ redshift catalog to reconstruct the large scale structure of the universe. The underlying density and velocity fields are reconstructed from the incomplete and noisy data out to a distance of $R = 8000Km/s$. Here we focus on mapping the Galactic ZOA, which in the case of the IRAS survey extends to $|b| \leq 5°$. At Gaussian smoothing of $1000Km/s$ the Great Attractor breaks down to three substructure whose peaks lie close to the Galactic plane within $270° \leq l \leq 330°$ and $3000 \leq R \leq 6000Km/s$. At the same resolution the Perseus-Pices is monolithic with no substructures.

1. Introduction

The mapping of the large scale structure (LSS) of the universe constitutes one of the main problems of modern cosmology. A key goal of this effort is the understanding of the origin of the streaming velocities of galaxies, and thereby the large scale dynamics of our local universe. This mapping relies primarily on redshift surveys of optically or IR selected galaxies (Saunders *et al.* 1991, Vogeley *et al.* 1992, Strauss *et al.* 1992, Fisher *et al.* 1993). The comparison of the velocities predicted from these surveys with actual observed velocities sheds light on such issues as structure formation, the 'biasing' problem (Dekel and Rees 1987) and the value of the cosmological parameters. Yet, redshift surveys provide only incomplete and noisy observational data on the underlying light and matter distribution. The Galaxy plane obscures a considerable fraction of the extragalactic sky, thus limiting the mapping to outside the so-called zone of avoidance (ZOA). The surveys are severely limited by the finite number of the observed galaxies which leads to considerable shot noise errors. The problem addressed here is that of the reconstruction of the underlying large scale density field from incomplete and noisy data.

There are two basic approaches to the problem which we label here as direct and indirect. The first one is the more commonly used in analyzing the LSS, and it assumes that the best estimator of the measured quantity is the data itself and its main effort is aimed at reducing the statistical errors. This is the case of the POTENT analysis of the velocity field (Bertschinger and Dekel 1989) and of the IRAS determination of the density field (Strauss *et al.* 1992). Another variant of this approach is the fit of the data by a parametric model, where the parameters are fitted by a maximum likelihood (or minimal χ^2 solution). Here, well known examples are the bulk flow and Great Attractor models of the velocity field

(Lynden-Bell *et al.* 1987). The drawback of this direct approach is that the best estimator of the underlying field one is measuring is always contaminated by the statistical noise and one cannot separate one from the other. A very different approach is presented by the Baeysian school of thought, where a model of the physical system at hand is assumed *a priori*. An estimator of the underlying field is then formulated within the framework of the *prior* model. The main advantage of this indirect approach is that it produces an estimator of the field **without** the noise. The price one pays, however, is the need to assume a *prior* model. Different variants of the Bayesian approach have been applied recently to the reconstruction of the LSS. Here we adopt the minimal variance estimator of the underlying field, also known as the Wiener filter (WF; Rybicki and Press 1992 and Zaroubi *et al.* 1994). The WF has been applied to the reconstruction of the angular galaxy distribution from the IRAS (2D) catalog (Lahav *et al.* 1994), the galaxy density field from the velocity potential (Ganon and Hoffman 1993), the IRAS $1.9Jy$ (Hoffman 1994), $1.2Jy$ redshift surveys (Fisher *et al.* 1994) and from the radial peculiar velocities (Stebbins 1994, whose estimator can be shown to be a WF), and the CMB temperature fluctuations from the COBE DMR data (Bunn *et al.* 1994). Here we shall present a reconstruction of the three dimensional density field from the IRAS $1.9Jy$ redshift survey (Strauss *et al.* 1992) which is based on the WF optimal estimator. This is part of a collaborative effort of analyzing the LSS by the WF method (Zaroubi *et al.* 1994). Here we focus on uncovering the LSS at the ZOA.

2.   Wiener Filter

Consider the case of a set of observations performed on an underlying field a_α yielding a data set $\{b_i\}$ ($i = 1.2., , , M$). Here we are interested in measurements that can be mathematically modeled by a linear convolution of the underlying field with a point spread function (PSF)

$$b_i = W_{i\alpha}a_\alpha + \epsilon_i, \tag{1}$$

where ϵ_i is the statistical error associated with the datum point i. For simplicity, and with no loss of generality, we assume α to be a discrete coordinate in some representation (such as Fourier, spherical harmonics or real space), $\alpha = 1, 2, , , N$. The PSF represents the response of the measuring device (or procedure) to the underlying field. Commonly it represents the blurring, or smoothing, introduced by the measurement. Here the notion of a PSF is extended to include also the theoretical procedure by which one field is expected to yield information on a different but related field. This is the case, for example, where a PSF is written to relate observations of the radial peculiar velocity field to the underlying density field.

Given the observable data one is interested in obtaining an optimal estimator of the underlying field. The approach adopted here assumes that the estimator depends linearly on the data set,

$$a_\alpha^{\mathrm{opt}} = F_{\alpha i}b_i. \tag{2}$$

The estimator is determined by the requirement that it minimizes the variance S from the actual field,

$$S = \left\langle |a_\alpha - a_\alpha^{\text{opt}}|^2 \right\rangle, \tag{3}$$

where $\left\langle ... \right\rangle$ denotes an ensemble average. Minimizing the variance with respect to $\mathbf{F}$ one finds the so-called WF,

$$F_{\alpha i} = \left\langle a_\alpha^* b_i \right\rangle \left\langle b_i^* b_j \right\rangle^{-1}. \tag{4}$$

Thus, the WF is the matrix product of the cross-correlation function of the data and the underlying field with the inverse of the auto-correlation matrix of the data. The optimal estimator of the underlying field is

$$a_\alpha^{\text{opt}} = \left\langle a_\alpha^* b_i \right\rangle \left\langle b_i^* b_j \right\rangle^{-1} b_j, \tag{5}$$

and the variance of the residual of the α-th degree of freedom is

$$\left\langle |r_\alpha|^2 \right\rangle = \left\langle |a_\alpha|^2 \right\rangle - \left\langle a_\alpha^* b_i \right\rangle \left\langle b_i^* b_j \right\rangle^{-1} \left\langle b_j^* a_\alpha \right\rangle \tag{6}$$

(note that here there is no summation over α). The error term ϵ_i is assumed to be statistically independent of the underlying field which yields

$$\left\langle a_\alpha^* b_i \right\rangle = W_{i\beta} \left\langle a_\alpha^* a_\beta \right\rangle \tag{7}$$

and

$$\left\langle b_i^* b_j \right\rangle = W_{i\beta} W_{j\beta'} \left\langle a_\beta^* a_{\beta'} \right\rangle + \sigma_i^2 \delta_{ij}. \tag{8}$$

Here, the errors covariance matrix is assumed, for the sake of simplicity, to be diagonal, $\left\langle \epsilon_i \epsilon_j \right\rangle = \sigma_i^2 \delta_{ij}$, however a general non-diagonal matrix can be easily used. The general functional form of the WF is a deconvolution of the PSF operating on the data, regularized by the ratio of $\left(prior / prior + noise \right)$.

The main important feature of the WF is its Baeysian nature, namely one assumes a *prior* model which specifies the statistical properties of the underlying field. The WF is optimal in the sense of minimal variance and the *prior* model is assumed in order to calculate the covariance matrix. Indeed, in the case where the random field is Gaussian the Bayesain most probable field is a WF estimator. The WF operates on the data to remove the shot noise and perform a deconvolution. Thus, it serves a dual purpose of estimation (of the field that has been observed) and prediction (of not directly observed fields). Within the framework of prediction the WF enables one to recover the structure hidden in the Galactic ZOA given an assumed *prior* model.

3. Data and Reconstruction

The reconstruction presented here is based on the density field of the IRAS 1.9Jy survey (Strauss *et al.* 1992) transformed to real space and smoothed with a Gaussian window of radius $R_s = 1200 Km/s$. The field is sampled within a sphere of a radius of $8000 Km/s$ on a Cartesian grid of $500 Km/s$ at a sampling rate of $1000 Km/s$. The sampling excludes a ZOA of $|b| < 5°$. The above criteria provide a set of 1896 constraints which are used for calculating the optimal underlying field within the entire $8000 Km/s$ sphere on the $500 Km/s$ grid. The underlying field is reconstructed with a $1000 Km/s$ Gaussian smoothing.

The *prior* assumed here is that structure grows by gravitational instability and a cold dark matter (CDM) power spectrum, which is determined by $\Omega_0 h$ and the σ_8 normalization (where Ω_0 is the cosmological density parameter and h is Hubble's constant in units of $100 Km/s/Mpc$). Here $\beta = \Omega_0 h$ is taken as a form factor to be determined from data, regardless of the value of the these cosmological parameters. The parameters assumed here are a CDM power spectrum normalized by $\sigma_8 = 0.7$ and a form factor $\beta = 0.2$, which constitute a good fit to the IRAS galaxy distribution (Fisher *et al.* 1993). The reconstructed field is not very sensitive to the value of these parameters, and generally we found the structure is reproducible for $0.2 \lesssim \beta \lesssim 0.5$ and $\sigma_8 \lesssim 1.0$.

4. Cosmography

The reconstructed density field is presented here by maps in Cartesian super-galactic coordinates and by full sky Aitoff projections in Galactic coordinates at different distances. As the main theme here is the reconstruction of the ZOA, the supergalactic coordinates maps focus on the $SGY = 0$ plane which almost coincides with the Galactic plane. The density and velocity fields at the $SGY = 0$ (*i.e.* roughly the Galactic) plane are presented in more details in Figs. 1(a,b). To gain a further insight to the three dimensional structure a series of angular maps at distances ranging from $R = 2000$ to $8000 Km/s$ (Figs. 2a-g), where fractional over density (δ) contours are plotted. The structure of our 'local' universe ($R \leq 8000 Km/s$) is dominated by the GA and PP overdensities surrounded by a region of an underdensity. The underdense regions percolate but they basically separate into two voids, the Local Void which extends from a distance of a few Mpcs out to the edge of the map around $30° \lesssim l \lesssim 90°$ and to very high and low latitudes, and a similar void around $l \approx 210°$.

The overdensity is dominated by the GA and PP structures with a filamentary structure which connects the two and includes the Local Group. The PP is the more dominant one in terms of its amplitude and monolithic structure that does not break into subclumps (at the $1000 Km/s$ resolution). The PP appears on the angular maps at $R = 3000 Km/s$ (Fig. 3c) and extends all the way to the edge of the the sample. It peaks at $R \approx 6000 Km/s, l = 150°$ and $b \approx -5°$ (Fig. 3e). Note that the densest part of the PP lies behind the ZOA and thus the reconstructed structure there is a 'prediction' of the WF. The GA is made of three substructures and seems to be located at the intersections of two filaments. The nearest structure is the well known Hydra/Centaurus supercluster at $R \approx (3 - 4) \times 10^3 Km/s, l \approx 300°$ and $b \approx 15°$ (Fig. 2a and Figs. 3b,c). The second density peak is located at $R = 6000 Km/s, l \approx 280°$ and $b \approx 5°$. This structure, which lies on the edge of the IRAS ZOA, was identified by Saunders *et al.* (1991) and more recently in an optical survey of the ZOA (Kraan-Korteweg and Woudt 1994).

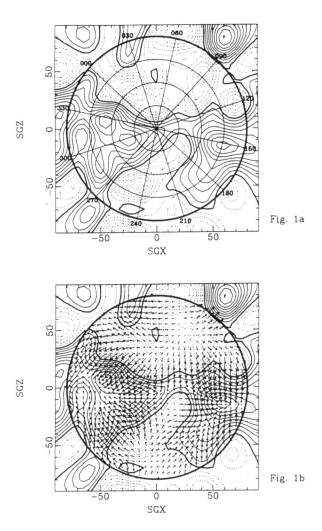

Figure 1.   Wiener Reconstruction of the density and velocity fields from the IRAS $1.9Jy$ redshift catalog in the SGY=0 plane. The data is obtain with $1200Km/s$ Gaussian smoothing and the field is reconstructed with $1000Km/s$ smoothing. The density field is sampled within a distance of $8000Km/s$, excluding the IRAS ZOA of $|b| < 5°$. The plotted angular rays almost coincides with the corresponding Galactic longitudes, and are given here for the sake of orientation. The circles correspond to distances of $2000, 4000, 6000$ and $8000Km/s$. The (fractional over) density is presented in Fig. 1a by contour plots with spacing of 0.1. The velocity field, superimposed on the δ contours, is given in Fig. 1b. The arrows represent the projection of the velocity vector at the $SGY = 0$ plane on that plane. The scaling of the length of arrows is arbitrary.

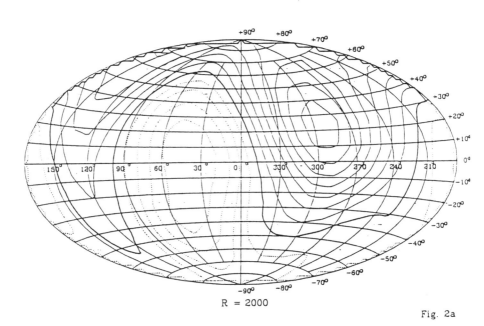

R = 2000

Fig. 2a

Figure 2.   The reconstructed density field is evaluated on spheres of
radii of 2000 (Fig. 2a), 3000 (Fig. 2b), 4000 (Fig. 2c), 5000 (Fig. 2d),
6000 (Fig. 2e), 7000 (Fig. 2f), and $8000 Km/s$ (Fig. 8g). Contour
of the δ-field at spacing of 0.1 are presented by Aitoff projection in
Galactic coordinates (l, b) at different distances.

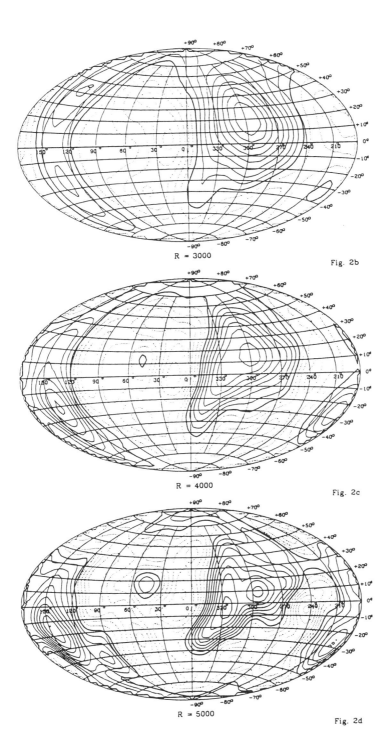

R = 3000

Fig. 2b

R = 4000

Fig. 2c

R = 5000

Fig. 2d

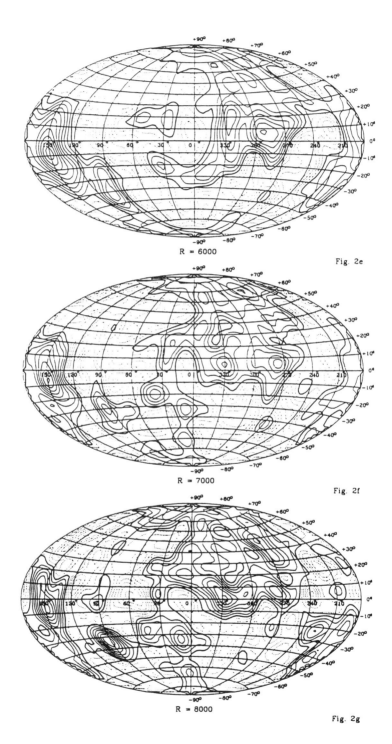

R = 6000

Fig. 2e

R = 7000

Fig. 2f

R = 8000

Fig. 2g

The most robust peak is the one that coincides with the GA location itself at $R \approx 5000 \, Km/s, l \approx 330°$ and $b \approx 0°$ (Fig. 2a and Fig. 3d). This peak is located just at the middle of the ZOA, and therefore is heavily obscured. Nevertheless, it coincides with a peak in the angular galaxy distribution of Kraan-Korteweg's (this volume) optical survey. Preliminary results of the redshifts reduction indicate that its redshift is ≈ 5000 (R. Kraan-Korteweg, private communication), and thus it coincides with the rich ACO cluster A3267.

The reconstructed velocity field is dominated by the overdensities at the PP and the GA complex on the one hand and by the underdensities (*i.e.* voids) described here. The flow converges towards the PP and GA and diverges from the voids. The POTENT velocity field (Dekel, Bertschinger and Faber 1990) is dominated by a coherent flow extending from the PP to the GA, a component which is induced by the tidal field. The lack of such a coherent flow, and the close correspondence with the density structure implies that the reconstruction does not fully recover the homogeneous solution of the Laplace equation, namely the tidal field. Note, however, that the reconstructed velocity field does not find a back infall towards the GA, but rather a flow away from it at $R \gtrsim 6000 \, Km/s$. This indicates that the WF predicts an overdensity roughly beyond the GA, outside of the region that is sampled here. This prediction is consistent with the raw IRAS structure.

A comparison with the raw 1.9Jy IRAS catalog (as presented by Kolatt, Dekel and Lahav 1994) and the richer 1.2Jy catalog (M. Davis, private communication) yields an overall agreement as far as the gross structure is concerned. Yet, these raw maps correct for the ZOA by an *ad hoc* mock sky procedure or a simple interpolation. These maps do not show the fine structure presented here and in particular the substructures of the GA.

5. Discussion

The comparison with the raw IRAS galaxy distribution shows the advantages of the WF. It cleans the raw maps from the shot noise and it predicts structure at unobserved regions. At the same computational effort the WF reconstructs also the velocity field, assuming the first order gravitational instability.

The study presented here is part of a more comprehensive effort of reconstructing the LSS. The method is now extended to reconstruct the perturbation field in real space from the galaxy distribution in redshift space. Thus providing a self consistent reconstruction tool which takes one from the purely observational plane to the primordial perturbation field. This will serve as a probe of the local cosmography, in particular the ZOA. The reconstructed perturbation field is currently used to set initial conditions for N-body simulations, to allow a dynamical modelling of the nonlinear regime.

Acknowledgments. I acknowledge my Wiener filter collaborators V. Bistolas, T. Bunn, K. Fisher, O. Lahav, D. Lynden-Bell, J. Silk and S. Zaroubi for their contribution to the present work and for many illuminating discussions. I am grateful to Renee Kraan-Korteweg for educating me the cosmography of our 'local' universe.

References

Bertschinger, E. and Dekel, A., 1989, ApJ, **336** , L5.

Bunn, E, Fisher, K.B., Hoffman, Y., Lahav, O., Silk, J., & Zaroubi, S., 1994, ApJ(submitted)

Dekel, A., Bertschinger, E., & Faber, S.M., 1990, ApJ, **364**, 349

Dekel, A. and Ress, M.J., 1987, Nature, **326**, 455

Fisher, K.B., Davis, M., Strauss, M.A., Yahil, A., & Huchra, J.P., 1993, ApJ, **402**, 42

Fisher, K.B., Lahav, O., Hoffman, Y., Lynden-Bell, D. & Zaroubi, S., 1994, MNRAS(submitted)

Ganon, G., and Hoffman, Y., 1993, ApJ, **415**, L5

Hoffman, Y., 1994, Proc. of the 9th IAP Conference *Cosmic Velocity Fields*, eds. F. Bouchet and M. Lachiéze-Rey, (Gif-sur-Yvette Cedex: Editions Frontiéres), pg. 357

Kolatt, T., Dekel, A., & Lahav, O. 1994 (preprint)

Lahav, O., Fisher, K.B., Hoffman, Y., Scharf, C.A., & Zaroubi, S., 1994, ApJ, **423**, L93

Kraan-Korteweg, R. and Woudt, P.A., 1994, Proc. of the 9th IAP Conference *Cosmic Velocity Fields*, eds. F. Bouchet and M. Lachiéze-Rey, (Gif-sur-Yvette Cedex: Editions Frontiéres), pg. 557

Lynden-Bell, D., Faber, S.M., Burstein, D., Davies, R.L., Dressler, A., Terlevich, R.J., and Wegner, G., 1987, ApJ, **326**, 19

Rybicki, G.B, & Press, W.H., 1992, ApJ, **398**, 169

Saunders, W., Frenk, C., Rowan-Robinson, M., Efstathiou, G., Lawrence, A., Kaiser, N., Ellis, R., Crawford, J., Xia, X.-Y., and Parry, I., 1991, Nature, **349**, 32

Stebbins, A., 1994, Proc. of the 9th IAP Conference *Cosmic Velocity Fields*, eds. F. Bouchet and M. Lachiéze-Rey, (Gif-sur-Yvette Cedex: Editions Frontiéres)

Strauss, M.A., Huchra, J.P., Davis, M., Yahil, A., Fisher, K.B., Tonry, J., 1992, ApJSuppl., **83**, 29

Vogeley, M.S., Park, C., Geller, M.J., and Huchra, J.P., 1992, ApJ, **391**, L5.

Yahil, A., Strauss, M.A., Davis, M., & Huchra, J.P., 1991, ApJ, **372**, 380.

Zaroubi, S., Hoffman, Y., Fisher, K.B., & Lahav, O., 1994 (preprint)

Discussion

D. Lynden-Bell: Given that every observer is concerned with big density contrasts of at least 1.5 but more likely a factor 3, why are you spending so much time and energy on a linearised theory?

Y. Hoffman: Lots of the observational work on the LSS focuses on linear structures (e.g. superclusters, streaming motions), and they should be properly modelled. Besides, linear theory is to be used as a starting point for non-linear modelling. Also, some of the ideas expressed here can be applied under the assumption of log-normal distribution.

Unveiling Large-Scale Structures Behind the Milky Way
ASP Conference Series, Vol. 67, 1994.
C. Balkowski and R. C. Kraan-Korteweg (eds.)

Modelling Large Scale Velocity Flows with Incomplete Sky Coverage

M. A. Hendry

Astronomy Centre, Univ. of Sussex, Brighton, UK

A. M. Newsam and J. F. L. Simmons

Dept. of Physics and Astronomy, Univ. of Glasgow, Glasgow, UK

1. Introduction

In recent years various methods (eg POTENT) have been developed to reconstruct the density and velocity fields from redshift surveys using redshift–independent distance indicators. It is well known (cf Lynden–Bell et al. 1988, Hendry et al. 1994) that any such method which groups together galaxies according to their *estimated* distance requires the application of a statistical correction – known as Malmquist correction (MC) – to each distance estimate, as a function of distance *and* of direction. In the treatment of Landy & Szalay (1992), however, reliable MCs can only be computed after first averaging over a large part – if not all – of the sky. Moreover the recovered velocity field will be subject to other sources of systematic error besides Malmquist bias: eg in POTENT additional biases arise from the use of large smoothing windows (Dekel 1994), and in 'attractor' models (cf Lynden–Bell et al. 1988, Han & Mould 1990) the best fit parameters are in general biased due to the non–linear form of the velocity field model.

2. Iterative Correction Procedure

We have developed a new, iterative, procedure designed to correct simultaneously for *all* sources of systematic error (due to distance errors, non–linear effects, density inhomogeneities and sampling gradients) in the reconstructed velocity field. Moreover, the bias corrections are directionally dependent, taking explicit account of, eg, incomplete coverage in the ZOA. The principle of the procedure may be outlined as follows.

1. Generate a large number of mock surveys, with randomly scattered distance estimates mimicking the real survey selection, but with redshifts equal to their observed values. Apply the reconstruction procedure to each survey and average. The resulting mean velocity field approximates to the true field plus *all* biases resulting from the reconstruction.

2. Repeat step (1) but with a prescribed model velocity field used to compute the mock redshifts. In this case the mean recovered field approximates to the prescribed model plus all reconstruction biases.

3. Since the model field is prescribed, compute directly the bias of the step (2) velocity field and subtract from the observed redshifts. (This assumes that the bias in steps (1) and (2) is approximately equal). Adopt the resulting velocity field as an improved model.

4. Repeat steps (2) and (3) iteratively until the residual biases are less than some acceptable level.

In the case of parametric model fitting the above procedure is applied in turn to each parameter of the velocity field model.

3. Application and Results

As an illustration we have applied the iterative correction procedure to the POTENT reconstruction of a realistic model velocity field model – containing both void and attractor features – using a mock galaxy catalogue derived from the Mathewson and Burstein Mark II surveys. We have compared the residual bias in our recovered velocity field with that of conventional POTENT reconstructions using both direct and inverse Tully–Fisher relations, and defining both homogeneous and inhomogeneous MCs, following the approach of Landy & Szalay (1992). We have adopted distance estimators with a scatter of $\sim 15\%$, and have included the effects of a sharp selection limit on apparent magnitude.

For this illustrative model velocity field we find that our iterative correction procedure produces residual biases which are more than 25% smaller than those obtained using all other combinations of distance estimators and Malmquist corrections which we investigated. Moreover, the recovery in the void region of the mock survey was considerably better than in any other case, supporting the idea that a directionally dependent bias correction is particularly important in sparse regions.

The principles behind the iterative method, and its practical application to POTENT reconstructions, are described in more detail in Newsam et al. (1994). In future work we will extend the method to the iterative correction of the reconstructed density field, with the aim of obtaining narrower confidence limits on the cosmological density parameter.

References

Dekel, A. 1994, ARA&A, to appear

Han, M., Mould, J. 1990, ApJ, 360, 448

Hendry, M.A., Simmons, J.F.L., Newsam, A.M. 1994, in 'Cosmic Velocity Fields', eds. Lachieze-Rey, M., Bouchet, F.R., Editions Frontieres, in press

Landy, S., Szalay, A. 1992, ApJ, 391, 494

Lynden–Bell, D., et al. 1988, ApJ, 326, 19

Newsam, A.M., Simmons, J.F.L., Hendry, M.A. 1994, A&A, submitted

Discussion

O. Lahav: How do you correct for Malmquist bias in your algorithm?

M. Hendry: The inhomogeneous Malmquist bias is removed from the peculiar velocity field **at the same time as** biases introduced by the smoothing window and sampling gradients in the velocity field. Our reasoning is that what one wishes is a peculiar velocity field free from all biases, of whatever source. The approach adopted in POTENT to remove inhomogeneous Malmquist bias from the distance estimates - even assuming that it does this perfectly - will only in turn give an unbiased velocity field if the smoothing windows are small. Moreover, Landy and Szalay's approach only works well after averaging over the sky: hence it deals with radial inhomogeneities, but not directional variations. For these reasons we wanted an algorithm which corrects the smoothed velocity field as a function of direction as well as radial distance, and removes all contributing biases, treating them all simultaneously.

J. Huchra: Do you think your technique is powerful enough, if presented with a good density field map plus a good density field derived from peculiar velocities, to derive variations in the biasing factor with position?

M. Hendry: I think we are probably a little way off being able to do that: the distance indicators are still too noisy. Perhaps in light of that, we should be trying to first address the problem of testing the universality of the distance indicators (indeed any environmental dependence of e.g. the Tully-Fisher zero point might well be physically linked to biasing!). To probe the positional dependence of b, we need to work on non-linear scales which are smaller than the smoothing radii currently needed by POTENT. Our method will do better than taking no account of the smoothing bias, but not yet well enough.

Part V

Radio Surveys

Unveiling Large-Scale Structures Behind the Milky Way
ASP Conference Series, Vol. 67, 1994.
C. Balkowski and R. C. Kraan-Korteweg (eds.)

Searching at 21 cm for Galaxies in the Zone of Avoidance

P. A. Henning

University of New Mexico, Department of Physics and Astronomy,
Albuquerque, NM 87131

Abstract. In the regions of highest optical obscuration and infrared confusion, only 21-cm emission can be used to find galaxies in the Zone of Avoidance. Past work and future prospects are reviewed.

1. Introduction

Diligent searches in the optical and infrared can narrow the Zone of Avoidance (ZOA), but the most opaque regions require searches using the 21-cm line of neutral hydrogen. Unlike at other wavelengths, the opacity of the Milky Way is never a problem at 21 cm (except for extragalactic HI emission near zero velocity, which would be masked by local hydrogen). At the lowest latitudes, HI searches are our only option for mapping large-scale structures.

In this contribution, the results of the feasibility study conducted with Dr. Frank Kerr, using the 300-ft telescope at Green Bank, and follow-up study with the Very Large Array will be discussed. Future prospects for full surveys, including one underway in the northern Zone of Avoidance, will also be presented.

The uniqueness of the search method presents an astrophysical opportunity: the study of a 21-cm-selected sample. Our present concept of a "normal galaxy" is based on optical catalogs, and to a lesser extent, on infrared (IR) compilations. Before the 300-ft search, there had never been such a large, blind 21-cm survey of galaxies. The sample produced by the feasibility study will be compared to various optical- and infrared- selected samples. Before a full survey of the ZOA is undertaken, it is vital to understand the sorts of objects we will find, how these differ from samples produced by surveys at other wavelengths, and the impact on our optically-prejudiced notion of normal galaxies.

2. The Feasibility Study

The pilot 21-cm survey for galaxies in the ZOA was conducted with the NRAO 300-ft telescope at Green Bank, WV. Five observing sessions during the period 1986-1988 covered 7200 search points. The beam width at 21 cm was about 10'. About 60% of the lines of sight were in the ZOA; the rest were located in regions of low obscuration as a control. The search points were not concentrated in any particular area, but were spread over the sky accessible to the 300-ft. (One exception is a concentrated search of the Pisces-Perseus supercluster and foreground void, discussed below.) The velocity range covered, −400 to 6800 km s^{-1} allowed the possibility of discovering new Local Group members, as well as spirals out to several thousand km s^{-1}. The 300-ft was an extremely sensitive instrument, allowing detection of dwarfs over a large volume, and normal spirals to the limit set by the spectrometer bandwidth. A more complete description of the survey method is provided by Henning (1992).

The survey yielded 37 extragalactic objects, and two interesting high negative velocity sources which we believe are associated with our own Galaxy. Only the sample of 37 clearly extragalactic objects will be discussed here. Nineteen of these appear in optical catalogs. As expected, most of the optically-known objects lie at high galactic latitude, but one, NGC 2377, is quite close to the Galactic Plane, at b = 3°. This illustrates the patchiness of the foreground obscuration. Of the full sample of 37 galaxies, 19 showed the typical two-horned or flat-topped HI profiles of unresolved spirals. But morphological classifications are shaky due to the random placement of the sources in the beam. Our rough classifications were accurate only about half of the time, judging from comparison with published morphological types of the optically-cataloged objects. Spirals far from the beam center tend to look one-horned, as do narrow-profile dwarf spirals. Therefore, all further study of the sample was based on follow-up observations with the Very Large Array (VLA) radio interferometer.

Interferometric observations were made for 25 galaxies, including all of the uncataloged objects and the weaker cataloged ones. The VLA was used in its most compact configuration, providing the highest sensitivity, but not allowing detailed mapping. From these observations, more accurate positions, HI masses, sizes, profile shapes, and linewidths were derived. The galaxies' HI masses range over two orders of magnitude, from 10^8 to 10^{10} $M_{\odot}$, small dwarfs to massive spirals. Figure 1 shows an example of an HI velocity profile of an optically-obscured spiral. It appears just like an optically-unobscured spiral! A_V makes no difference whatsoever to HI.

Armed with accurate positions, we consulted the Infrared Astronomical Satellite (IRAS) Point Source Catalog to investigate any IR counterparts. Twelve of the 37 objects are associated with IR sources of appropriate color for galaxies. Ten of these are optically cataloged, and the other two appear on the Palomar Sky Survey. Why such a poor IR performance finding the hidden galaxies? There are two reasons. First, the IR sky is terribly confused at low latitudes, so our sources near the Galactic Plane were masked by Galactic IR emission. Even inspection of the more sensitive IRAS co-adds does not help pick out extragalactic sources at the very lowest latitudes. Second, IRAS is biased against dwarfs and dwarf spirals. The optically-cataloged dwarf and dwarf spirals in our sample were not listed in the Point Source Catalog, even though they were well away from the obscuration. A more detailed comparison of HI versus IR methods of finding galaxies in the ZOA will be presented later.

3.   A Warning for Future Surveys

During the course of the survey, we came across six narrow-lined sources near the Galactic Plane, all with velocities of about 4500 km s^{-1}. We were intrigued by the possibility of the discovery of a previously unknown cluster behind the Milky Way. However, the sources' velocities were too similar to be physical. We conjectured that these objects were not extragalactic sources of HI emission, but

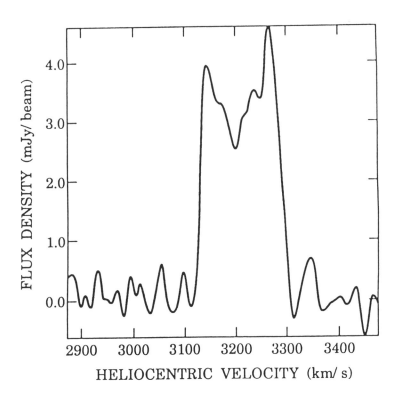

Figure 1.   Profile of a spiral in the Zone of Avoidance. Note the two-horned shape typical of spirals. The galaxy lies at b = 3°. There is nothing but stars and dust visible on the Palomar Sky Survey prints at this location.

local high-quantum-number hydrogen recombination line emission. H167α at a rest frequency of 1399.368 MHz mimics HI redshifted to 4500 km s^{-1}. To test this theory, we looked for H166α at these positions, and found it in each case. Future surveyors should beware of narrow-lined sources near 4500 km s^{-1}, and check for other recombination lines at the appropriate frequencies.

4.   Results of the HI Survey in Two Cosmic Voids

As a special portion of the feasibility study, we surveyed two cosmic voids. One, the Pisces-Perseus foreground void, is in an area of low obscuration, away from the Galactic Plane. The other, the Local Void, lies in a large portion of the ZOA. In each case one galaxy was discovered in HI. A full description of the Pisces-Perseus search and follow-up optical work is given by Henning and Kerr (1989). In both cases the object found was of low HI mass ($\approx 10^8 M_\odot$), low optical surface brightness, and late type. The two galaxies rank numbers 23 and 24 of the 25 galaxies studied with the VLA. While it is a dicey business to make generalizations based on a sample of two, it is intriguing that the objects are so similar, and may be giving us a clue about the sorts of objects residing in voids.

The cosmologically important question is: do these sorts of low surface brightness objects which are detectable in HI **fill** the voids? In the case of the Pisces-Perseus void we had too few search points (628) to draw conclusions about this. If the number density of HI-selected galaxies were not a function of direction, we would expect only 3 galaxies in this void. Detecting 1 is not statistically different from expectation. However, we searched over 1500 lines of sight in the Local Void and found only 1 galaxy. In this case we should have found 10 galaxies if they were uniformly distributed. We conclude that the Local Void is underdense in HI galaxies at the 3-σ level. Low surface brightness, HI-rich galaxies do not fill voids.

5.   Comparison of the HI-Selected Sample with Optical- and IR-Selected Samples

Using the 21-cm line of HI to search for galaxies is a new technique, and it is important to understand the sorts of objects which are found this way. Will a 21-cm-selected sample consist of the same sorts of galaxies which are found in optical- or IR-selected samples, or will they differ? This is important to know when comparing various studies in the ZOA, and also is of general astrophysical interest.

Because a large portion of the HI-selected sample lies in optically opaque regions, we are forced to consider only the HI properties of the sample *versus* HI properties of the various comparison samples. For this analysis, the distribution of HI linewidths is considered, a sensitive indicator of galaxy morphological type. The possible bias against broad-lined galaxies has been considered, and does not appear to be important. A fuller discussion of this analysis will appear in a future publication (Henning 1994). First, optically-selected samples are considered.

5.1. A Sample of Isolated Galaxies

The HI in galaxies which reside in dense cluster environments often shows signs of disruption, either by tidal interaction with nearby galaxies, or by ram pressure stripping by intercluster media. In order to study intrinsic HI properties of galaxies as a function of morphological type, one must look at isolated objects. Galaxies which appear in both the Uppsala General Catalogue of Galaxies (Nilson 1973; UGC) and the Catalog of Isolated Galaxies (Karachentseva 1973) were studied at 21 cm by Haynes and Giovanelli (1984). They observed 324 galaxies, and detected 288 in HI. Early type galaxies (earlier than Sa) make up 5% of their sample, and 14% are later than Sc. Using the Kolmogorov-Smirnov statistical test to compare linewidth distributions, we determine at the 98% confidence level that the HI-selected sample could not have been drawn from the same population as the optically-selected sample of isolated galaxies.

5.2. A Magnitude-Limited Sample

The next sample considered is a magnitude-limited sample. This type of collection is often used in large optical redshift surveys, so it is important to understand how an HI-selected sample might differ. To date, there has been no optical magnitude-limited sample which has been uniformly studied in HI. For this study, we chose all galaxies in the Revised Shapley Ames catalog (Sandage and Tammann 1981; RSA) brighter than $B_T = 12.0$ and $|b|$ greater than 40° and which have published HI information. We took the linewidths from the literature, or measured them directly from the published profiles, if necessary. This sample of 262 galaxies consists predominantly of Sc's (35%), only 8% later than Sc, and 9% earlier than Sa. Comparing the linewidth distributions of this sample and our HI-selected sample, we find that the two samples are drawn from different populations of galaxies at the 99% confidence level.

5.3. Dwarf and Other Low Surface Brightness Galaxies

We've seen that the HI-selected sample is not like collections of high surface brightness galaxies. The next sample considered is a sample of dwarf and other low surface brightness galaxies, studied by Schneider et al. (1990). They observed all of the galaxies in the UGC which are listed as "dwarf", "irregular", "Sd-m", or later, and which are accessible to the Arecibo telescope. The resulting sample of 574 detected galaxies has a linewidth distribution which is statistically indistinguishable from that of the HI-selected sample. So while the HI search did not discover a new class of previously unknown objects, it does turn up galaxies which are difficult to study optically, and are under-represented in optical catalogs.

5.4. Low-Latitude IRAS-Selected Galaxies

Finally, we consider the populations of galaxies which are discovered through IRAS searches in the ZOA and the HI search method. A Cornell/Arecibo group (Dow et al. 1988; Lu et al. 1990) compiled a list of galaxy candidates in the latitude range $|b| = 2° - 16°$, and made pointed observations in HI. Of the list of 371 candidates, 26% were detected at 21 cm and of those, 37% were not previously known as galaxies. The linewidth distribution of the galaxies found in this way is more unlike that of the HI-selected sample than any of the other samples we considered. The IRAS sample consists only of large spirals, as discussed in an earlier section. The two methods of searching the ZOA are complementary, and it is important to realize that they trace **different populations of galaxies.**

6. Clustering Properties and Space Density of the HI-Selected Sample

We have determined that the HI-selected sample is not made up of the same sorts of objects that are usually used to trace large-scale structures. Because blind 21-cm searches are rarely done, the relative distributions of HI-selected galaxies and optical samples are not well known. For our sample, we can ask if the HI galaxies out of the ZOA lie along structures delineated by optically-selected samples. In particular, we considered the magnitude-limited CfA redshift survey (Huchra *et al.* 1983) and a volume-limited survey of dwarf and other low surface brightness galaxies. Analysis by Thuan *et al.* (1991) indicates that the high- and low-surface brightness galaxies lie along the same three-dimensional structures. We have eleven HI-selected galaxies in regions covered by these surveys. Each of the eleven lies along structures traced by the high- and low-surface brightness samples. This result is consistent with the paucity of HI galaxies in the Local Void (discussed in Section 4).

Finally, we consider the space densities of galaxies of various HI masses. In particular, when relying on optical catalogs, are we missing a large population of low optical surface brightness, HI-rich objects? To answer this, we have constructed an HI mass function, analogous to optical luminosity functions. Figure 2 shows the number density of galaxies as a function of M_{HI}, weighted inversely by the volume in which the galaxies could have been detected. The search sensitivity is also shown. This sample is larger than any previously available for such an analysis, and shows an upturn at low M_{HI}. However, we do not see the large population of very low mass (10^7 to 10^8 $M_\odot$) low surface brightness dwarfs predicted by the model of dwarf galaxy formation developed by Tyson and Scalo (1988). Our search sensitivity would have allowed detection of this population.

7. Future Prospects in the North and South

The time is ripe for a full 21-cm survey of the ZOA. There is no other way to map large scale structure in the regions of highest optical obscuration, and infrared confusion. A full survey has begun in the northern ZOA: the Dwingeloo Obscured Galaxies Survey (DOGS). An international group of astronomers in the Netherlands, United Kingdom, and the United States has been conducting tests of the telescope and data reduction system, and plans to survey the entire ZOA accessible from Dwingeloo. The intended coverage is between galactic longitudes 30° and 220°, and latitudes −5° and +5°, and heliocentric velocities out to 4000 km s^{-1}. As of the time of writing, blind searching has just begun. The intention is to map the distribution of spirals, and nearby dwarf galaxies. The survey will take a couple of years, depending on final coverage and integration time used.

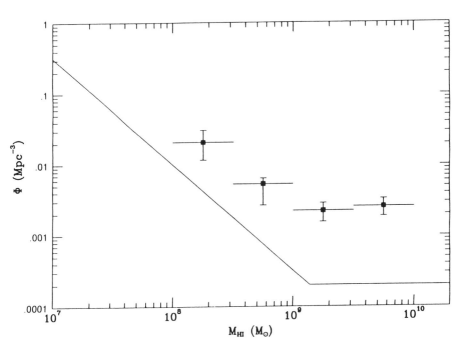

Figure 2.   The number density of galaxies binned in M_{HI}, weighted inversely by the volume over which they could have been detected. The sensitivity of the search is shown by the solid line.

In the southern sky, new hardware plans may make a full survey of the ZOA feasible. Investigation is underway of a multi-beam 21-cm receiver system for the Parkes telescope in Australia. This system would cut down the time required for a full HI survey by almost an order of magnitude over what is necessary with present technology. The system would open up the 21-cm sky, not just in the ZOA, but could be used all over the southern celestial sphere to determine the definitive HI mass function, to study gas-rich objects, to survey Galactic and Magellanic HI, and to conduct a host of other projects. More details of this proposed undertaking are given in the poster presentation by the ATNF group at this conference.

References

Dow, M.W., Lu N.Y., Houck, J.R., Salpeter, E.E., and Lewis, B.M. 1988, ApJ, 324, L51

Haynes, M.P., and Giovanelli, R. 1984, AJ, 89, 758

Henning, P.A. 1992, ApJS, 78, 365

Henning, P.A. 1994, in prep

Henning, P.A., and Kerr, F.J. 1989, ApJ, 347, L1

Huchra, J., Davis, M., Latham, D., and Tonry, J. 1983, ApJS, 52, 89

Karachentseva, V.E. 1973, Catalog of Isolated Galaxies, Comm. Spec. Astrophys. Obs. 8

Lu, N.Y., Dow, M.W., Houck, J.R., Salpeter, E.E., and Lewis, B.M. 1990, ApJ, 357, 388

Nilson, P. 1973, Uppsala General Catalogue of Galaxies, Uppsala Astr. Obs. Ann., Vol. 6 (UGC)

Sandage, A., and Tammann, G.A. 1981, A Revised Shapley-Ames Catalog of Bright Galaxies, Carnegie Inst. of Washington, Pub. 635 (RSA)

Schneider, S.E., Thuan, T.X., Magri, C., and Wadiak, J.E. 1990, ApJS, 72, 245

Thuan, T.X., Alimi, J.-M., Gott, J.R., Schneider, S.E. 1991, ApJ, 370, 25

Tyson, N.D., and Scalo, J.M. 1988, ApJ, 329, 618

Discussion

H. van Woerden: (1) In the zone of avoidance you find a galaxy population different from that at higher latitudes. Don't you find in addition at least some galaxies of that high-b population (among your optically uncatalogued galaxies)?
(2) Could you give the detection limit for the Dwingeloo survey now being undertaken? I expect it will depend on redshift, velocity width and possibly other factors.

P. Henning: (1) Yes, there are some massive spirals among our uncatalogued galaxies. In fact, our most massive spiral ($M_{HI} \sim 10^{10}$ $M_\odot$) is uncatalogued, with $b = 0°.6$.
(2) I would like to hold off giving definite numbers until the results of our tests are digested.

N. Lu: Is your upper limits on HI dwarfs with $M_{HI} \lesssim 10^8$ lower than that in Hoffman, Lu and Salpeter (1993)?

P. Henning: Yes, we can set more stringent limits on the number density of these dwarfs because we surveyed a much larger volume.

J. Huchra: With the ~ 10 K galaxies in the CfA survey in complete regions we have been able to calculate the best luminosity function for dwarf galaxies so far - just because of sheer numbers. The optical galaxy luminosity function is rising again at faint absolute magnitudes just like your HI mass result. Zwicky was right!

O. Lahav: Your HI mass function looks very different (with 'opposite curvature') from optical (Schechter) and IRAS luminosity functions. Does the flat part of your function correspond to the flat part of a Schechter function, and the rise at the HI faint-end has no counterpart in the optical luminosity function? Do you probe here a different population in HI? What is known about the relation between HI and optical luminosities for galaxies with both measured properties? Could you form a bi-variate HI-optical function?

P. Henning: We are probing a different population in HI. The dim end of the Schechter function is dominated by dE's, while our low HI mass objects are low optical surface brightness, gas-rich dwarfs. Frank Briggs has constructed a theoretical HI mass function based on the Schechter function and the relation between HI mass and optical luminosity for dwarfs and spirals derived by Fisher and Tully, $M_{HI}/M_{HI}^* \approx (L/L^*)^{0.9}$. Our HI mass function looks quite different from his formulation, which is based on optically-selected objects. I would like to extend the work to search at 21-cm out of the ZOA. Then we can look at the HI mass optical luminosity question.

Recent Radio Continuum Surveys of the Southern Milky Way and a Future MultiBeam HI Survey

Ronald T. Stewart

ATNF,CSIRO,PO Box 2121,Epping,NSW,Australia.

Abstract. Recent radio continuum surveys of the southern Galactic plane include the following:

(1) An .843 GHz survey of the plane by MOST with 43 arcsec resolution for $|b| > 1.5°$ and $255 < \ell < 355°$ by Whiteoak et al. (1989), Whiteoak (1993) and from $355° < \ell < 360°$ by Gray (1993).

(2) A 4.8 GHz survey of the plane by the Parkes 64m telescope with 5 arcmin resolution as part of the Parkes-MIT-NRAO (PMN) southern sky survey (Griffiths and Wright 1993). This has been followed by AT observations of some 8000 compact sources with $S < 70$ mJy to obtain source sizes and positions down to 1 arcsec. Because of confusion near the plane the AT observations are restricted to $|b| < 2°$ (Tasker et al. 1993).

(3) A 2.3 GHz survey of the plane from $|b| < 5°$ and $238° < \ell < 365°$ by Duncan, Stewart, Haynes and Jones (1994).

Also a 21 cm HI multibeam survey of the southern sky and Galactic plane is being planned for the Parkes telescope beginning in 1996 (Stavely-Smith et al. 1994).

From (1) Whiteoak (1993, 1992) has found 1800 compact sources between $257° < \ell < 270°$ and $282° < \ell < 299°$ of which about 80 % are probably extragalactic.

From (3) Duncan et al (1994) find that about 80% of their compact sources are correlated with PMN compact sources.

No doubt the catalogues ensuing from these surveys will be a valuable source of informatiom fot future studies of galaxies in the Southern Milky Way.

References

Duncan, A.R., Stewart, R.T., Haynes, R.F. & Jones, K.L. 1994, Proc ASA, submitted

Gray, A. 1993 , PhD Thesis, University of Sydney

Griffith, M., Wright, A.E., Burke, B., & Ekers, R.D. 1993, ApJS, in press

Staveley-Smith, L., Wilson, W.E., Bird, T.S., Sinclair, M.W., & Ekers,R.D. 1994, Workshop on Multi-Feed Systems for Radio Telescpoes, Tucson, Arizona May 16-18, 1994 (ASP conference series)

Tasker, N.J., Wright, A.E., and 5 others, 1993, Proc ASA, 10 (4), 320

Whiteoak, J.B.Z. 1993, PhD Thesis, University of Sydney

Whiteoak, J.B. 1992, A&A, 262, 251

Whiteoak, J.B.Z., Large, M. I., Cram, L.E., & Piestrzynski B., Proc ASA, 8, 176

Discussion

H. van Woerden: (1) You mention 1400 galaxies between longitudes $257°$ and $270°$ found in the MOST survey. Is this classification as galaxies based on radio morphology from maps or on spectral data?
(2) The Tornado source at $\ell \sim 357°$, $b \sim 0°$ reminds me of three things : a) the Arc source on the other side of the Galactic Centre ($\ell \sim 0°2$?) which also has parallel curved filaments; b) at least one planetary nebula which has corkscrew morphology; c) the parallel filaments in the North Polar Spur. I am not suggesting the Tornado is a planetary nebula. I think the arrangement of filaments must be due to a strong magnetic field.

R. Stewart: (1) As far as I know these compact 843 MHz MOST sources were classified as extragalactic sources, on the basis that they were not coincident with IRAS PSC sources (with colours typical of galactic sources such as HII regions or planetary nebula). At 843 MHz non-thermal sources predominate thermal HII regions because of steeper spectral index. Very few compact non-thermal sources are supernova remnants so I think it is very likely that most compact radio sources at 843 MHz are radio galaxies. It is not possible at present to check the spectral index of the MOST sources because there is no comparable survey at another frequency.
(2) The Tornado is too bright and too large to be a planetary nebula.

E. Meurs: Following Hugo van Woerden's question, did all of your radio sources have an IRAS counterpart?

R. Stewart: No, only about 1%.

T. Yamada: On the Molonglo survey: did Whiteoak cross correlate all the detected sources with the IRAS PSC? There is a tight correlation between 60 μm and 21-cm (or 6 cm) radio continuum for galaxies. Your detection limit in Jy (which wavelength?) could be ~ 0.5 Jy at $60\mu m$ and thus as deep as IRAS. Once you cross-correlate the detected sources with IRAS, you can identify galaxy candidates from the ratio between $60\mu m$ and the radio continuum.

R. Stewart: Whiteoak did correlate radio and IRAS PSC and only found 1% correlation.

Unveiling Large-Scale Structures Behind the Milky Way
ASP Conference Series, Vol. 67, 1994.
C. Balkowski and R. C. Kraan-Korteweg (eds.)

HI Studies Related to the Optical Thickness of Spiral Galaxies

L. Bottinelli and L. Gouguenheim
Observatoire de Paris, F-92195 Meudon Cedex

G. Paturel
Observatoire de Lyon, F-69230 Saint Genis Laval

P. Teerikorpi
Tuorla Observatory, SF-21500 Piikkiö

Abstract. We point out the implication of our work on the Tully-Fisher distance indicator, using an angular size limited sample of more than 5 000 spiral galaxies, on the opaqueness problem. We conclude that the inclination dependent behaviour of these galaxies is consistent with a high optical thickness at the surface brightness limit 25 magnitude per square second, in any case significantly larger than the classical value of about 0.2.

1. Introduction

The optical thickness of galaxy disks has been discussed in several recent papers with contradictory conclusions ranging from optically thin (Huizinga and van Albada, 1990 ; Byun, 1993) through "moderately optically thick" (Burstein et al., 1991) to "opaque"edges (Valentijn, 1990 ; Choloniewski, 1991) of the spiral disks. One reason contributing to the unclear situation, is the problem how to treat correctly the data in statistical studies (Burstein et al., 1991).

Our contact with the present problem comes from a practical goal of how to correct magnitudes and diameters to zero inclination. This knowledge, which is closely related to the question of optical thickness, is needed in our programme (Paturel et al., 1990) for measuring the kinematics of the local universe using the Tully-Fisher (TF) distance indicator together with an angular size limited sample of 5 171 spiral galaxies.

2. Method

From observed $\log V_M$ (V_M = maximum rotation velocity), we predict diameter D_{25} and magnitude B for spiral galaxies of different inclinations ($\log R_{25}$) using the TF relations together with kinematical distances derived from a Virgo infall model. In the ideal case, the observed and predicted parameters, $\log [D_{obs}/D_{pred}]$ and $B_{obs} - B_{pred}$ vs. $\log R_{25}$ diagrams should directly give the corrections. In practice, such an approach may be influenced by inclination dependent Malmquist bias. Hence, we first constructed unbiased samples, using the concept

217

of normalized distance d_n which has proved so useful in studies of H_o (Bottinelli et al., 1986). We add to the normalization formula an inclination term, in order to guarantee that at all inclinations the samples are free from the Malmquist bias, when one cuts the sample at sufficiently small d_n.

This approach overcomes the problems pointed out by Burstein et al. and by Choloniewski for the methods not utilizing distance information. An important advantage is the possibility to construct sufficiently large unbiased magnitude limited subsamples - apparent magnitudes B_T are known for almost 90% of the sample - and hence to test simultaneously the diameter and magnitude effects.

In order to intercompare the results from diameters and magnitudes, we adopt a simple disk + spherical bulge model for the galaxies, and assume that the bulge is free from inclination dependence, while the surface brightness of the disk may change with the viewing angle.

Writing in a linear approximation (R stands for R_{25}):

$$\log [D_{obs}/D_{pred}] = C \log R \qquad (1)$$

and denoting $k = L_{bulge}/L_{tot}$ and $K_{25} = dlogD/d\mu$, the corresponding magnitude change within the 25 magnitude isophote will be:

$$m_{obs} - m_{pred} = -2,5 \log[k + (1-k)R^{2C(0.2/K_{25}+1)-1}] \qquad (2)$$

In principle, the bulge fraction k and the slope of the surface brightness profile K_{25} are known for different galaxy types, and the coefficient C is the only free parameter. For the model to be realistic, C should be consistently given by both the diameter and magnitude behaviour.

We cut from the sample very face-on and (for the diameter test) very edge-on galaxies (allowing $0.07<logR<0.8$), because face-on galaxies have large errors in $logV_M$ and edge-on galaxies have a systematic effect in their diameters (Huyzinga and van Albada, 1990; Burstein et al., 1991).

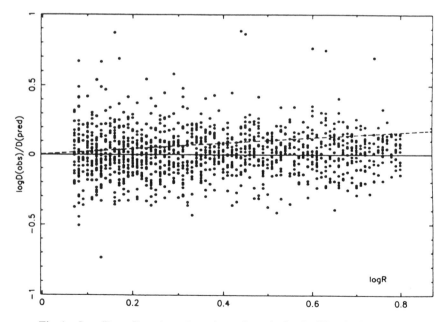

Fig.1. $Log(D_{obs}/D_{pred})$ against the axis ratio logR. The dashed line corresponds to the "classical" correction C=0.20.

3. Results

We show first that for Sb - Scd types there is very small, if any, diameter effect up to $\log R = 0.8$. The smaller unbiased sample already gave us an indication for $C \approx 0$, which means that we can check the value of C using the whole angular size limited sample in the diameter test: with $C \approx 0$, the Malmquist bias at each $\log R$ would be quite similar. In Fig. 1 one notes the impressively constant mean value up to $\log R = 0.8$, a formal least-squares solution giving $C=0.04\pm0.02$.

The complementary test consists in checking whether such a small value of C is consistent with the magnitude behaviour. Here it is important to use the unbiased subsample of a magnitude limited sample $(m<12.5)$ which is different for each value of C (entering the formula for the normalized distance). The predicted magnitude effect is quite sensitive to C. Fig. 2 shows the result for $C=0.0$, 0.04, and 0.1, using the value $K_{25}=0.1$ (from Fouqué and Paturel, 1985) and the values of the bulge luminosity fraction (from Simien and de Vaucouleurs, 1986). $C=0.1$ is clearly too large, while $C \approx 0.04$ gives the best fit, in agreement with the result from the diameter effect.

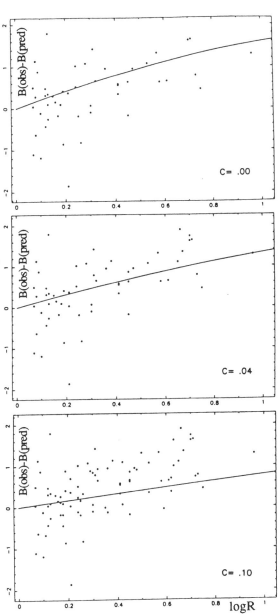

Fig.2. B_{obs} - B_{pred} against $\log R$ for three different unbiased magnitude limited sub-samples. The curves give the corresponding expectation from Eq.2.

Note that the magnitude effect is less than expected for a pure disk (C=0 would imply the slope =2.5), roughly by the amount required by the known contribution from the bulge. We concluded that the diameters and the magnitudes alike indicate a very small value of C, and a correspondingly high value of the optical thickness τ. But how high? A rough estimate comes from the model where stars and dust are uniformly mixed at the disk edge (Disney et al., 1989). Ignoring any scattered light, the radiation transfer problem can be solved for each inclination and the coefficient C=dlogD/dlogR becomes

$$C=2.5K_{25}R\tau \exp(-\tau R)/[1-\exp(-\tau)] \tag{3}$$

which is not constant but may be integrated to give the total change of the size when the galaxy is turned away from the face-on position:

$$\log D/D_o=2.5K_{25}\log\{[1-\exp(\tau R)]/[1-\exp(-\tau)]\} \tag{4}$$

where D_o denotes the face-on diameter and D the diameter for a galaxy seen with the axis ratio R.

It is not possible to determine τ accurately (for $\tau >1.0$ its value is far too sensible to small changes in the observed C). However, the small values of C suggests typically $\tau >0.8$, or at least "moderately optically thick" Sb - Scd disks. $\tau <0.5$ is clearly excluded, together with the "classical" inclination corrections corresponding to C≈0.2.

References

Bottinelli, L. Gouguenheim, L., Paturel, G., Teerikorpi, P. 1986, A&A 156, 157
Burstein, D., Haynes, M.P., Faber, S.M. 1991, Nature 353, 515
Byun, Y.I. 1993, Publ. Astron. Soc. Pac. 105, 993
Choloniewski, J. 1991, MNRAS 250, 486
Disney, M.J., Davies, J.L., Phillips, S. 1989, MNRAS 239, 939
Fouqué, P., Paturel, G. 1985, A&A 150, 192
Huizinga, G.E., Van Albada, T.S. 1990, MNRAS 254, 677
Paturel, G., Bottinelli, L., Fouqué, P., Garnier, R., Gouguenheim, L., Teerikorpi, P. 1990, The Messenger, n°62, 8
Simien, F., Vaucouleurs, G. de 1986, ApJ 302, 564
Valentijn, E.A. 1990, Nature, 346, 153

Discussion

N. Lu: How do you correct D for inclination effects in calibrating your $(D - W_V)$ TF relation?

L. Gouguenheim: We use an iterative method, starting from a given value of opaqueness. Such an iteration has to be made also for selecting the unbiased plateau, particularly when using the magnitude Tully-Fisher relationship.

H. van Woerden: In spiral galaxies, the average HI surface densities (projected to face-on) are often about 1×10^{21} atoms cm^{-2}, but not much higher (see e.g. Warmels 1988, *Astron. Astrophys. Suppl.* **72**, 427 and 453; Broeils 1992, thesis Groningen). Local values may be much higher, but note that linear resolutions in HI maps of galaxies are generally not better than 100-1000 pc.

Part VI

IRAS Selected Samples

Unveiling Large-Scale Structures Behind the Milky Way
ASP Conference Series, Vol. 67, 1994.
C. Balkowski and R. C. Kraan-Korteweg (eds.)

An HI Search for IRAS Galaxies in the Galactic Plane

L. Bottinelli, L. Gouguenheim, M. Loulergue, J.M. Martin, G. Theureau
Observatoire de Paris, section de Meudon F-92195 Meudon Cedex

G. Paturel
Observatoire de Lyon, F-69230 Saint Genis Laval

Abstract. 74 new galaxies were discovered in the Zone of Avoidance, from their 21-cm line emission, starting from FIR flux densities and colours criteria. They are on the mean distant and very luminous in the FIR.

They are used, together with new 21-cm line velocity measurements, to study various concentrations of galaxies in the Zone of Avoidance.

1. Selection of the candidates

A list of 669 candidates was selected from the IRAS point source catalogue (PSC, 1988), using flux densities and FIR colours criteria. The area covered is bounded by $|b| < 20°$ and $\delta \geq -39°$, and is outside the Arecibo declination range. It results the following longitude coverage:

$$0 \leq l \leq 30° \; ; \; 350° \leq l \leq 360° \; ; \; 85° \leq l \leq 180° \; ; \; 210° \leq l \leq 255°$$

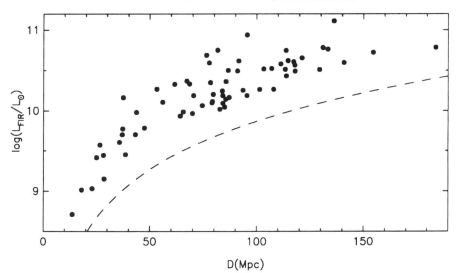

Figure 1. Logarithm of the FIR luminosity L_{FIR} vs. the distance ($H_0 = 75$ km s^{-1} Mpc^{-1}) for our newly identified IRAS galaxies. The dashed curve corresponds to our flux densities limits

2. HI Observations

We used the Nançay radio telescope in velocity search mode in the velocity range 400 - 10 000 km s^{-1} (Martin et al., 1990).

Up to now 300 sources were observed in the 21-cm line; 74 were detected, 7 of them being without any optical counterpart. The unpublished radial velocities are given in Table 1.

Fig. 1 shows the FIR luminosity, derived according to Lonsdale et al. (1985), vs. distance and the FIR luminosity bias resulting from our selection. It is quite striking that most of these new galaxies are rather distant high FIR luminosity galaxies, instead of nearby ones, with moderate FIR luminosities.

Table 1. Parameters of the newly discovered IRAS galaxies.

IRAS name	l	b	V	σ	IRAS name	l	b	V	σ
1	2	3	4	5	1	2	3	4	5
00059+5514	116.8	-6.9	5154	5	05393+5359	158.1	12.5	8124	9
01213+5555	127.5	-6.4	5408	7	05475+4837	163.5	11.0	5913	9
01583+6807	129.4	6.4	3578	7	05480+4316	168.2	8.4	13240	7
02297+6351	133.8	3.4	4202	10	05523-1330	218.9	-18.6	6728	7
02471+5430	139.7	-4.2	4534	11	06036-0032	208.2	-10.3	2048	11
03076+6712	135.9	8.1	3153	14	06102+4823	165.5	14.2	9341	
03108+6441	137.4	6.1	2391	6	06234-1654	225.4	-13.2	6214	6
03189+5828	141.6	1.4	2320	2	06274-2829	236.8	-17.0	7110	15
03231+3721	154.0	-15.8	5546	9	06324-3135	240.2	-17.2	6984	9
03277+6755	137.1	9.8	1333	3	06370+3623	178.8	13.6	7644	12
03354+6633	138.5	9.1	1496	1	06555-0516	218.4	-1.0	2723	2
03409+4534	151.7	-7.2	8104	20	07155-2215	235.7	-4.6	2816	7
03495+3957	156.6	-10.6	5600	15	07170-0058	217.1	5.8	9949	9
03527+5051	150.0	-1.9	4675	10	07510-2209	239.7	2.7	7072	12
03571+6022	144.3	5.8	7851	11	08175-1433	236.6	12.0	6015	20
04043+4541	154.8	-4.5	4470	6	08254-3512	254.7	1.8	2033	9
04116+5506	149.3	3.2	4849	9	08567-1811	245.2	17.6	2082	6
04162+6557	142.1	11.3	8270	25	16400-2034	358.6	16.4	1206	10
04167+4118	159.5	-6.2	6061	7	16434-1447	4.0	19.3	8218	12
04175+4035	160.1	-6.6	6057	5	16585-1129	9.0	18.1	7302	5
04235+5638	149.4	5.4	4744	4	16591-1654	4.5	15.0	8529	26
04287+4239	160.1	-3.6	9255	9	17134-1539	7.5	12.8	8530	11
04312+4008	162.3	-5.0	6141	12	19348-0619	32.6	-13.1	3074	16
04322+4736	156.9	0.2	6469	7	19387+0145	40.4	-10.2	5897	7
04379+3843	164.2	-5.0	8099	16	21292+4801	92.2	-2.3	3789	4
04436+4849	157.3	2.4	2913	4	21441+4454	92.0	-6.3	7597	12
05000+5109	157.1	5.9	9500	5	21524+4130	91.0	-9.9	5497	13
05134+5811	152.5	11.6	5320	10	22114+5015	99.0	-4.9	8555	27
05206+4217	166.4	3.6	11098	13	22282+7506	114.2	15.0	2445	7
05251+5544	155.5	11.6	7291	13	22284+3825	95.0	-16.4	693	1
05287+4228	167.1	4.9	6161	5	22325+6810	110.8	8.9	4355	6
05326+5016	160.8	9.7	5906	10	22493+4944	104.0	-8.4	9996	16
05327+0251	201.4	-15.5	5819	8	22547+5336	106.4	-5.3	4844	6
05376+5057	160.6	10.7	5970	12					

Explanation of columns: (1) IRAS name; (2) Galactic longitude, in degrees; (3) Galactic latitude, in degrees; (4) Systemic radial velocity ($c\Delta\lambda/\lambda_o$, in km s^{-1}) from 21-cm line; (5) Mean error on velocity.

3. Discussion

We give here preliminary results on the distribution of the galaxies in the (l,b) plane, adding the velocity data available either in LEDA (Durand et al., this workshop) or from our (ESO + Nançay) key-programmes (Paturel et al., 1990).

3.1. Region 80° ≤ l ≤ 180°

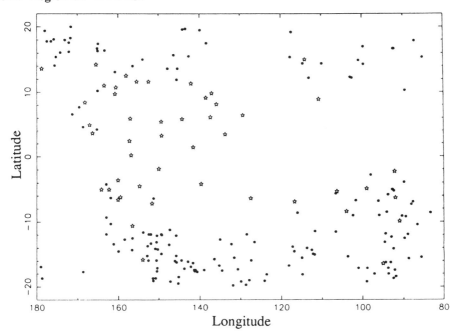

Figure 2 : Galactic distribution of the galaxies in the longitude range 80° ≤ l ≤ 180°. The stars stand for the newly discovered IRAS galaxies

We find new members of the Perseus-Pisces filament (Burns and Owen, 1979, Focardy et al., 1984, Hauschildt, 1987, Balkowski et al., 1988, Chamaraux et al., 1990, Maurogordato et al., 1991); we confirm the concentration of galaxies in the velocity range 4 000 - 5 000 km s^{-1}, suggested by Huchra et al. (1977) and by Weinberger (1980); we find new low velocity (2 400; 2 800; 3 300 km s^{-1}) galaxy groups near (l=100°, b=+15°).

3.2. Region 180° ≤ l ≤ 280°

Our data confirm the existence of several groups of galaxies in the Puppis (l ≈ 240°) concentration of galaxies (Kraan-Korteweg & Huchtmeier, 1992; Yamada et al., 1993), whose some properties are listed in Table 2.

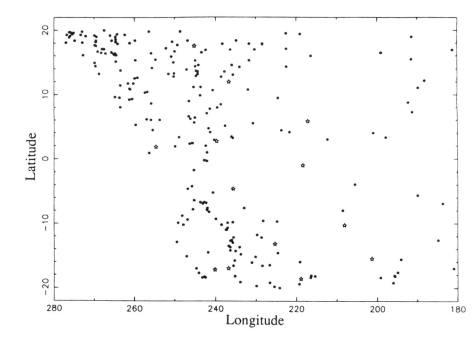

Figure 3. Galactic distribution of the galaxies in the longitude range
180° ≤ l ≤ 280°. The stars stand for the newly discovered IRAS galaxies

Table 2. Groups of galaxies in the Puppis region.

Group	l	b	n	\<V\>	σ
1	2	3	4	5	6
1	245	-10	14	2053	159
2	250	- 9	11	2823	126
3	245	+ 3	11	1611	89
4	235	+11	10	4603	225
5	240	+12	7	5750	143
6	250	+13	15	1883	184
7	255	+16	23	2481	125

Explanation of columns: (1) Group number; (2) and (3) Galactic longitude and
latitude in degrees; (4) Number of group members; (5) and (6) Mean radial
velocity and dispersion in km s^{-1}.

3.3. Sagittarius-Ophiuchus concentration

We find also new members in this concentration of galaxies in the region
(l = 359°; b = 8°) (Wakamatsu & Malkan, 1981; Johnston et al., 1981; Djorgovski
et al., 1990), clustered around two different mean velocities: 8 600 and 11 400
km s⁻¹.

References

Balkowski, C., Cayatte, V., Chamaraux, P., Fontanelli, P. 1988, The World of
   Galaxies, p. 420, H.G. Corwin and L. Bottinelli ed., Springer-Verlag
Burns, J.O., Owens, F.N. 1979, AJ 84, 1478
Chamaraux, P., Cayatte, V., Balkowski, C., Fontanelli, P. 1990, A&A 229, 340
Djorgovski, S., Thompson, D.J., de Carvalho, R.R., Mould, J.R. 1990, AJ 100, 599
Focardi, P., Marano, B., Vettolani, G. 1984, A&A 136, 178
Hauschildt, M. 1987, A&A 184, 43
Huchra, J., Hoessel, J., Elias, J. 1977, AJ 82, 674
IRAS Point Source Catalogue 1988, Joint IRAS Science Working Group
   (Washington, DC : US GPO) (PSC)
Johnston, M.D., Bradt, H.V., Doxsey, R.E., Margon, B., Marshall, F.E., Schwartz,
   D.A. 1981, ApJ 245, 799
Kraan-Korteweg, R., Huchtmeier, W.K. 1992, A&A 266, 150
Lonsdale, C.J., Helou, G., Good, J.C., Rice, W. 1985, Catalogued Galaxies and
   Quasars observed in the IRAS survey, Jet Propulsion Laboratory, Pasadena
Maurogordato S., Proust, D., Balkowski, C. 1991, A&A 246, 39
Martin, J.M., Bottinelli, L., Dennefeld, M., Fouqué, P., Gouguenheim, L., Paturel,
   G. 1990, A&A 235, 41
Paturel, G., Bottinelli, L., Fouqué, G., Garnier, R., Gouguenheim, L., Teerikorpi, P.
   1990, The Messenger, n°62, 8
Wakamatsu, K., Malkan, M. 1981, PASJ 33, 57
Weinberger, R. 1980, A&AS 40, 123
Yamada, T., Yakata, T., Djamaluddin, T., Tomita, A., Aoki, K., Takeda A., Saito,
   M. 1993, MNRAS 262, 79

IRAS GALAXIES IN THE PERSEUS
SUPERCLUSTER NORTH-EAST EXTENSION

Maren Hauschildt Purves

*California Institute of Technology, Division of Physics, Mathematics
and Astronomy, Caltech Submillimeter Observatory, 1059 Kilauea
Avenue, Hilo, HI 96270, USA*

Abstract. A sample of tentative IRAS galaxies in the galactic zone
of avoidance north-east of the Perseus Supercluster is presented. These
galaxy candidates have been found from positional coincidence with optical sources. From the literature as well as from HI line observations,
radial velocities are available for part of the sample, confirming these
objects as galaxies beyond any doubt. More than half the galaxies for
which radial velocities are known in this sample are part of the Perseus
Supercluster.

Introduction

Several methods have been used to find galaxies in the galactic zone of avoidance, some of which had been previously used for galaxies at higher galactic
latitudes, like optical searches on the Palomar Observatory Sky Survey and infrared searches of the IRAS Point Source Catalog (IRAS Catalogs, 1985, hereafter
PSC). Other methods, like "blind" HI searches (see Henning, 1992), have been
developed specifically to find galaxies in the galactic zone of avoidance.

The sample discussed here results from a combined optical/IRAS approach.

Sample Definition

This sample originates from a search on the Palomar Observatory Sky Survey
E prints for galaxies in the area north-east of the Perseus Supercluster, between
galactic longitudes of about $140°$ and $170°$ and latitudes $-10°$ and $+10°$, but
following the edges of the Palomar Observatory Sky Survey prints rather than
galactic latitudes and longitudes. This optical search resulted in some 10,000
objects, with coordinates accurate to about $0.1 - 0.2'$. The list of optical objects
was then compared with the PSC, using a cut-off distance of $1'$ between IRAS
and optical positions, independent of optical diameter and quality of the IRAS
position. The only condition placed on the IRAS sources in this was that they
have to be detected at either 60 or 100μ, i.e. not applying any (other) color
criteria, no 100μ flux limitation and retaining extended sources in the sample.

This comparison resulted in a list of 243 optical objects associated with 231
IRAS sources.

Most objects found by Lonsdale et al. (1985) were found here too, differences are discussed in a forthcoming paper.

Figure 1 shows the distribution of differences between optical and IRAS position versus optical diameter, uncorrected for galactic extinction. Filled triangles are objects confirmed as galaxies by redshift measurements, open triangles are candidate objects yet to be confirmed, whether listed in catalogs of galaxies or not.

Quite in contrast to what might be expected, the difference between optical and IRAS positions for larger objects tends to be smaller.This in spite of the fact that they are not necessarily stronger IRAS sources which might have more accurate IRAS positions.

A Selection Effect Found For Small Diameter Optical Objects

Figure 1 shows a small gap in the data at the origin and extending upwards and to the right, characterized roughly by a line where the separation is about twice the diameter. The factor of two can easily be explained by the effect of galactic extinction on galaxy diameters. Applying a correction of 2 magnitudes roughly doubles the diameter (see Hauschildt, 1987 or Cameron, 1990).

For the larger objects, all corrected diameters are larger than the difference between optical and IRAS coordinates, not only due to the limit on optical – IRAS associations. In fact most separations are smaller than the uncorrected diameters. In contrast to that, the smaller objects are separated by the gap, below which the coordinate difference is larger than the corrected diameter. The smaller objects seem to consist of two entirely different populations. Possible explanations for the objects below the gap are:

1. Optical coordinates are less accurate for smaller objects

2. If smaller (optical) objects are weaker IRAS sources, the IRAS coordinates could be less accurate

3. If larger (optical) objects are stronger IRAS sources, a possible statistically distributed population of unrelated IRAS sources might have merged into those stronger IRAS sources

4. Most of the smaller optical objects are not galaxies but images of superposed faint stars, plate flaws, wisps of extended galactic stuff. All of these are objects we don't want when looking for galaxies and which aren't necessarily associated with the IRAS sources.

While there is probably some contribution involved for all of these reasons, the last one seems to be the most likely and the main contributor here. As a consequence, in the following, the objects below the gap are discarded in spite of the fact that two of them are now known from radial velocity measurements to be galaxies.

This leaves a sample of 205 objects associated with 198 IRAS sources (some are multiple galaxies). At this point, radial velocities are known for 83 of 198 or 89 of 205, i.e. slightly over 40%.

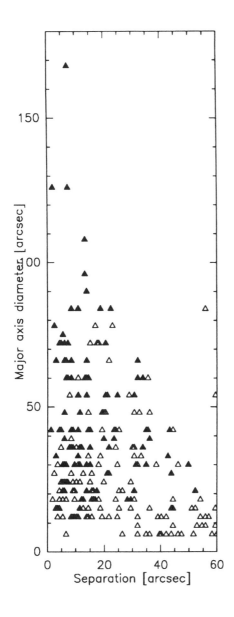

Figure 1.   Difference between optical and IRAS coordinates (separation) versus optical diameter of the objects. Optical diameters are not
corrected for galactic extinction. Filled triangles are galaxies for which
radial velocities are known, open triangles are candidate objects which
have yet to be confirmed

IRAS Colors - In General And In Particular

The selection here involves optical as well as IRAS sources, and as this was meant to catch as many galaxies as possible no IRAS color criteria have been applied other than that the object in question had to be detected at 60μ or 100μ. The usual criteria of a 12μ detection has been dropped as it would leave only 23 objects in the sample, but even the "relaxed" color criteria for normal galaxies by Martin et al. (1990), F25/F60 < 0.5, F60/F100 < 1 and F100 < 10 Jy would have excluded most objects from the sample. This of course implies that not all galaxies in the sample are necessarily normal galaxies.

One of the criteria dropped here is the requirement for a detection at 25μ. Only 49 of the objects in the sample are 25μ detections, with 26 of them known to be galaxies from radial velocities, i.e. a confirmation rate similar to the whole sample, but cutting away 75% of objects in the sample, which was not the objective.

Another one of the criteria dropped is the 100μ flux restriction. There are 18 objects with 100μ fluxes in excess of 10 Jy in this sample. 5 of them have measured extragalactic radial velocities, the highest flux being 23 Jy at 5755 km/s.

Figure 2 shows the IRAS flux ratios, not logarithms, of the objects in the sample that were detected at 60μ. Upper limits are shown as negative numbers, to retain the information that one (or two) of the fluxes involved are upper limits rather than detections. The upper right quadrant contains the ones detected in all 3 wave-bands, the lower right contains the ones that have been detected at 60 and 100μ, the upper left the ones detected at 25 and 60μ and the lower left the ones detected only at 60μ. Three of the four quadrants contain objects that we know for certain are galaxies. The exception is the quadrant containing objects detected at 25 and 60μ but not at 100, but this quadrant is largely empty anyway. Filled circles are objects for which radial velocities are known that confirm them as galaxies. The sample by Martin et al. (1990) is shown as crosses for comparison, independent of whether radial velocities are known or not.

The 5 objects in the sample that were detected only at 100μ are excluded from figure 2, but 2 of them are confirmed galaxies.

Implications For The Perseus Supercluster

Figure 3 shows the distribution of the objects in this sample in right ascension and declination, open triangles being objects yet to be confirmed as galaxies and filled triangles being confirmed galaxies.

Figure 3 contains only objects in this sample. These objects shown in right ascension and declination alone cannot provide any more conclusive evidence for a continuation of the Perseus Supercluster through the galactic zone of avoidance, not even if taken together with data available for other galaxies in this area. A comparison of the radial velocity distribution inside an area $5°$ to either side of a line connecting Abell 426 at $3^h12^m +41°$ and Abell 569 at $7^h04^m +48°$ to the galaxies outside this area is shown in figure 4. Figure 4a shows the galax-

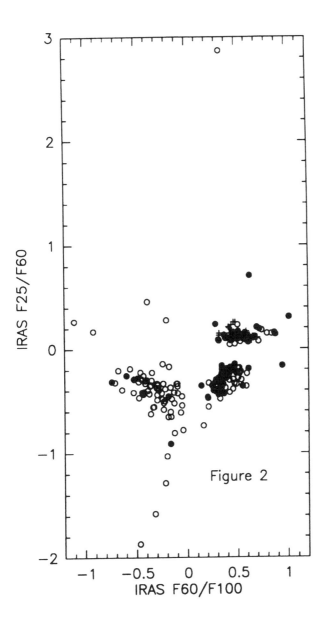

Figure 2.   F25/F60 versus F60/F100 flux ratios. Upper limits are shown as negative numbers. Objects not detected at 60μ are omitted from the figure. For further explanations see text.

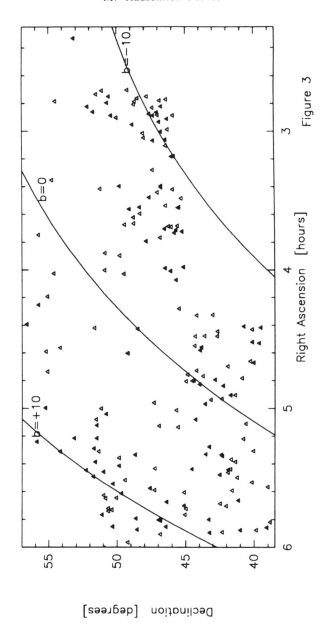

Figure 3.   Distribution of the objects in this sample in right ascension
and declination. Filled triangles are galaxies for which radial velocities
are known, open triangles are candidate objects yet to be confirmed.
Galactic latitudes b = -10°, 0° and +10° are indicated

ies within 5° of the axis connecting Abell 426 and Abell 569 whereas figure 4b shows the galaxies outside this area.

Figure 4 shows evidence from the galaxies in this sample alone that the Perseus Supercluster extends across the galactic zone of avoidance. The majority of the galaxies close to the axis are at Perseus Supercluster radial velocities whereas galaxies further away from this axis do not show the same marked concentration in radial velocity.

References

L.M. Cameron, 1990, A&A **233**, 16

M. Hauschildt, 1987, A&A **184**, 43

P.A. Henning, 1992, ApJS **78**, 365

IRAS Catalogs, 1985, eds. C.A. Beichman, G. Neugebauer, H.J. Habing, P.E. Clegg, T.J. Chester; Washington D.C., Government Printing Office (PSC)

C.J. Lonsdale, G. Helou, J.C Good, W. Rice, 1985 *Cataloged Galaxies and Quasars Observed in the IRAS Survey*, Jet Propulsion Laboratory, Pasadena

J.M. Martin, L. Bottinelli, M. Dennefeld, L. Gouguenheim, 1990 A&A **235**, 117

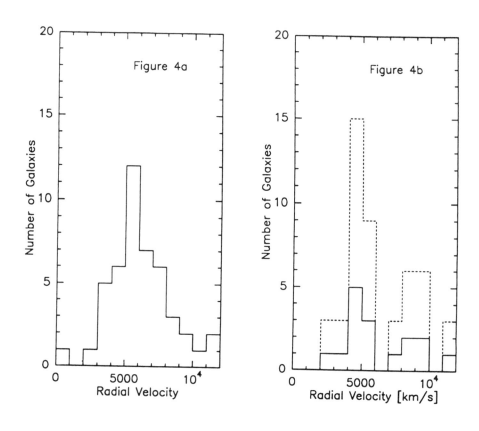

Figure 4.   Radial velocity distribution of the galaxies in this sample for which radial velocities are known. Figure 4a shows the galaxies within +/- 5° of the line connecting Abell 426 and Abell 569, figure 4b the galaxies outside this area. The dotted line in figure 4b shows the same histogram re-scaled for the same number of galaxies as in figure 4a.

Large-Scale Structures in the Highly Obscured Orion-Taurus Region

Nanyao Y. Lu

IPAC/JPL, MS100-22, California Institute of Technology, Pasadena, CA 91125

W. Freudling

Space Telescope-European Coordinating Facility – ESO
Karl-Schwarzschild-Strasse 2
D-W8046 Garching bei Muenchen, Germany

Abstract.
    A sample of IRAS galaxy candidates were observed in HI 21cm for galaxy identification and redshift measurements. This generated a uniform sample of IRAS galaxies in the area of $2^h < \alpha < 10^h$ and $0° < \delta < 36°$ which crosses the Zone of Avoidance (ZOA) and includes most of the heavily obscured Orion-Taurus region. The representative galaxy distribution from our resulting galaxy sample provides a view on large-scale structure in this area unbiased by Galactic reddening.
    The main results are: (1) the possibility of a nearby, nearly optically hidden, very rich galaxy concentration in this region can probably be ruled out. (2) The main part of Pisces-Perseus supercluster is limited to $\alpha < 3^h$ in our survey region by giant voids between 3^h and 4^h. (3) There are excessive galaxies around velocities of ~ 5000 and $\sim 8500\,\mathrm{km\ s^{-1}}$, respectively. The latter "wall" appears to gradually diffuse out after it enters the ZOA from the northern Galactic hemisphere.

1. Introduction

Galaxies have been identified optically in the Zone of Avoidance (ZOA) with or without using IRAS positions (e.g., Böhm-Vitense 1956, Fitzferald 1974, Dodd & Brand 1976, Weinberger 1980, Kraan-Korteweg 1989, Saitō et al. 1990 & 91, Yamada et al. 1993, Takata et al. 1994, and papers in this book). While optically generated galaxy samples are quite uniform at high Galactic latitudes, they become highly nonuniform near the Galactic plane ($|b| \lesssim 30°$) and in regions of excessive and patchy reddening, making it difficult and, perhaps, ambiguous to interpret the observed galaxy distribution in the ZOA. Blind HI 21cm surveys (e.g., Kerr & Henning 1987) would produce a uniform sample, but is very time consuming.
    IRAS Point Source Catalog (Version 2, 1988; hereafter PSC) is, on the other hand, fairly complete above a few degrees from the Galactic plane. A follow-up HI 21cm observation on far-infrared (FIR) selected PSC candidates allows one

TABLE 1. Infrared Selection Crteria

Parameter	Criteria
PSC flux density at 100μm:	$1.5 < f(100) < 8$ Jy
PSC flux quality at 100μm:	moderate or high
PSC flux density ratios:	$f(25)/f(100) < 0.50$
	$f(12)/f(100) < 0.17$
	$1.13 < f(100)/f(60) < 4.00$
PSC correlation coefficients (CC):	$CC(60) \geq 0.98$; $CC(100) \geq 0.98$

$f(x)$ is the PSC flux density at x μm, with $x = 12, 25, 60$ or 100.

to generate a galaxy sample which is free from Galactic extinction (Lu et al. 1990) and the resulting galaxy sample offers an unbiased view of the large-scale structure in and across the ZOA.

We have applied the above extinction-free approach to the region bounded by $2^h < \alpha < 10^h$ and $0° < \delta < 36°$. The part of the region with $b < 0°$, which contains most of the highly obscured Orion-Taurus region, is of several interests: (i) The region was always masked out by previous redshift surveys; (ii) The "local velocity anomaly" of the Local Group is pointing in (or near) this region (Faber & Burstein 1988), a possible hint for a (unknown) nearby rich galaxy concentration in this region; (iii) The main part of Pisces-Perseus (hereafter PP) supercluster is limited to $\alpha < 3^h$ in optical in this region. Despite extensive studies on the PP region (e.g., Haynes & Giovanelli 1988), no specific redshift survey has been done in the current region to see if the main part of the PP supercluster extends to $\alpha > 3^h$. (iv) The "Great Wall" (Geller & Huchra 1989) optically "disappears" in the ZOA from the northern Galactic hemisphere. It is important to see if it ends there or extends to the Orion-Taurus region, in order to better determine the characteristic length of this largest known coherent structure. The part of our survey region with $b \gtrsim 10°$, which has much less obscuration, was selected for the purpose of some statistical comparison.

2. The Sample Selection and Observations

2.1. The Sample Selection

In heavily obscured regions, Galactic source density in the PSC is also high. Following Lu et al. (1990), we use a set of selection criteria, given in Table 1, which screens out more than 75% Galactic sources, while still includes more than 50% IRAS galaxies (Lu et al. 1990; also see Meurs in this book). Applying our criteria to (optically) cataloged galaxies in the PSC at high Galactic latitudes, we indeed selected at least 50% at all redshifts. The adopted limiting flux density of 1.5 Jy at 100μm is equivalent to about 0.7 Jy at 60μm.

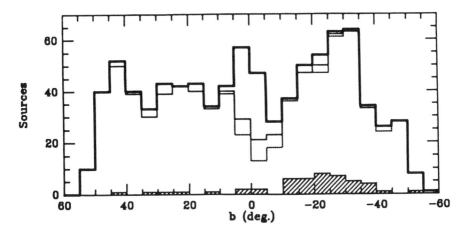

Figure 1.   Plot showing our survey completeness along Galactic latitude

2.2. Observations

The selected sample of 876 sources can be divided into three parts: (i) 462 (53%) "cataloged galaxies", namely, those noted in PSC to be associated with one or more optical galaxy catalogs; (ii) 47 (5%) "Galactic sources" (as marked in the PSC); and (iii) 367 (42%) pure IRAS sources. Our main observational goal is to observe every source in parts (ii) and (iii), which were not known to be a galaxy to us by the time of our actual observations. To this end, we searched literature for identified galaxies in our sample. Both optical and radio identifications were accepted.

The 21cm observations were made with the 305m radio telescope of the Arecibo Observatory. A candidate source was usually first observed for the low-velocity range (-400 to 8200 km s^{-1}). If no galaxy signal was detected, another spectrum was taken for the high-velocity range (7800 to $16,400$ km s^{-1}). Each resulting spectrum has a velocity resolution of ~ 16 km s^{-1} and a typical rms noise of 1.4 mJy (after Hanning smoothing). The HI 21cm survey of Lu *et al.* (1990),with a velocity coverage up to 9000 km s^{-1}, is technically very similar to the current one, but mainly limited to $|b| < 16°$. It was therefore integrated to our observations over the low-velocity range.

3. Survey Completeness

Our survey completeness up to some heliocentric velocity, v_h, in our sample can be defined as the number percentage of those sample sources which are either known to be galaxies or observed (up to v_h) in one of the above mentioned HI surveys. Figure 1 shows the distribution of our sample sources along Galactic latitude (the thick solid line). The distribution of those sample sources which are either known to be galaxies or observed over our low (low + high) velocity coverage is shown by the thin solid (dotted) line in Fig. 3. Therefore, our survey

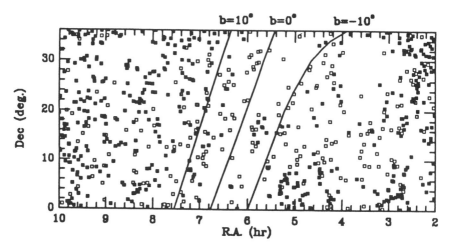

Figure 2.   Distribution of Sample Galaxies in Equatorial Coordinates

is nearly complete up to $\sim 16,000$ km s^{-1}for $|b| > 10°$. For $|b| < 10°$, the completeness is somewhat lower, but up to 8000 km s^{-1}, we are more than 80% complete down to $|b| = 5°$. The hatched histogram is the distributions of known "non-galaxies" up to our limiting velocity of 16, 000 km s^{-1}, which comprise less than 15% percent of our sample at $b < -10°$ and less than 6% at $b > 10°$.

Our HI detections should be quite complete up to $v_h \approx 8000$ km s^{-1}(Lu et $al.$ 1990). At higher velocities, we may lose some galaxies with low HI flux density in face-on position.

4.   Results

4.1.   Overall Galaxy Distribution

The sky distribution of all 717 known galaxies is shown in Figure 2, where we have symbolically differentiate the cataloged galaxies (filled squares) from the pure IRAS galaxies (open squares). The percentages of pure IRAS galaxies in four equal-area bins along R.A. are, respectively, 33% (for $2^h - 4^h$), 53%, 43%, and 20%. The variation of this ratio should mainly reflect the pattern of the variation of Galactic extinction A_v. A_v is quite small for the last bin of $8^h < \alpha < 10^h$ (Burstein & Heiles 1982), the difference between pure and cataloged IRAS galaxies there reflects largely the true selection effect. As we shall see in Figure 4, the majority of pure IRAS galaxies (27 out of 29) in this R.A. bin are at velocities higher than 8000 km s^{-1}, while most of the cataloged galaxies are below 10, 000 km s^{-1}. These results show: (i) the existing optical catalogs of galaxies could be 33% (= 53% − 20%) incomplete in the Orion-Taurus region, and (ii) in regions of small Galactic extinction, our selection does not select many optically unknown galaxies below 8000 km s^{-1}.

83% (594) of the identified galaxies have heliocentric velocities v_h available to us. We reduced v_h to v_c, the velocity with respect to the centroid of the Local Group, using $v_c = v_h + 300\sin(l)cos(b)$ (km s^{-1}) where l and b are Galactic

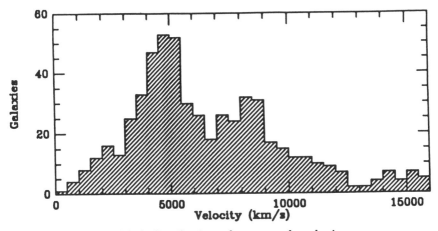

Figure 3.   Redshift distribution of our sample galaxies.

longitude and latitudes, respectively. Figure 3 shows the overall distributions of v_c up to 16,000 km s^{-1} for sample galaxies. For a comparison, a sample selected with $f(60) > 0.7$ Jy (see § 2.1.) from a uniformly distributed galaxy population characterized by one of the available 60μm luminosity functions (e.g., Yahil et al. 1991) would have a redshift distribution which peaks around 4000 km s^{-1}. The overall pattern in Figure 3 can be characterized by two "valleys" (or voids) and two "peaks", superposed on the underlying smooth distribution. The two apparent voids are one below $\sim$ 4000 km s^{-1} and one around $\sim$ 7000 km s^{-1}. The first peak has a factor of $\sim$ 2 overdensity at $\sim$ 5000 km s^{-1}. Galaxy concentrations around this velocity were also observed in various directions in and around our survey region (e.g., Focardi, Marano, & Vettolani 1984; Hayschildt 1987; Dow et al. 1988; Haynes & Giovanelli 1988; Maurogordato 1991; Giovanelli & Haynes 1993; Takata et al. 1994; Seeberger, Huchtmeier & Weinberger 1994). As shown below, this "wall" of galaxies around 5000 km s^{-1} is due to a number of superclusters between 4000 and 6000 km s^{-1} and relative deficiencies of galaxies off the wall in the two voids. The secondary peak centers at about 8500 km s^{-1}, a velocity corresponding to that of the Great Wall (Geller & Huchra 1989). As we shall see below, our data suggests that the Great Wall diffuses out in the ZOA.

4.2. Galaxy Distribution Over Smaller Scales

Galaxy distributions over scales much smaller than our survey scale can be visually illustrated by the cone diagrams in Figure 4, where we plot v_c versus R.A. for the whole survey area (panel [a]) and for three declination zones of 12 degrees each (panels [a] to [c]). We only show galaxies with $v_c < 12,000$ km s^{-1} above which our data appear sparse in such cone diagrams. The area between the two dashed lines in each diagram indicates $|b| < 5°$, where our observations are not quite complete yet.

A few known superclusters can be easily identified in Figure 4: the "head" of PP supercluster (Haynes & Giovanelli 1988) in Figure 4d ($v_c \sim 5000$ km s^{-1},

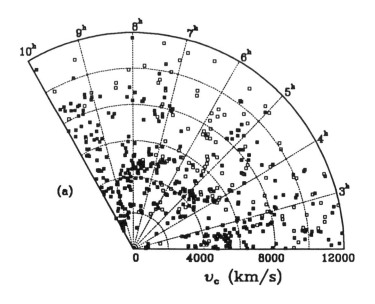

v_c (km/s)

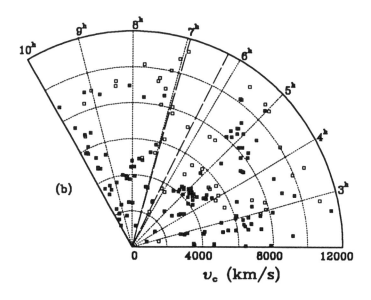

v_c (km/s)

Figure 4.   Cone diagrams of velocity versus R.A. for various zones of declinations δ: (a) the whole survey area, and (b) $0° < \delta < 12°$ on this page; (c) $12° < \delta < 24°$, and (d) $24° < \delta < 36°$ on the next page. The symbols are the same as in Figure 3.

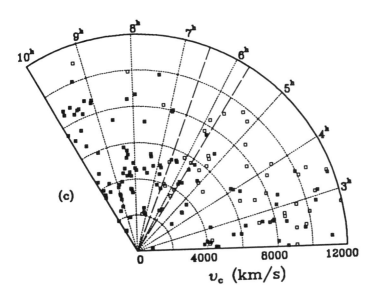

(c)

v_c (km/s)

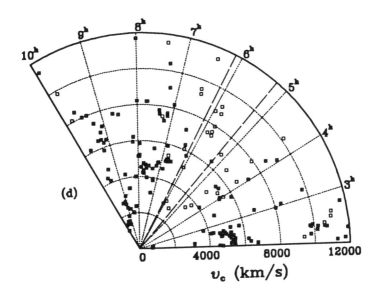

(d)

v_c (km/s)

$2^h < \alpha < 3^h$), the N1600 supercluster (Saunders *et al.* 1991) in Figure 4c ($v_c \sim 4400$, $\alpha \sim 5^h$), the Gemini filament (Focardi, Marano, & Vettolani 1984) in Figures 4c and 4d ($4000 < v_c < 6000$, $6^h < \alpha < 8^h$, part of the Great Wall (Geller & Huchra 1989) in Figures 4b and 4c ($8000 < v_c < 10,000$, $\alpha > 7^h$). Our statistical analysis also shows that the concentration around $\alpha \sim 2.4^h$ and $v_c \sim 9100$ km s^{-1}in Figure 4d is probably fairly high in galaxy density and that there exists a significant supercluster at $\alpha \sim 6^h$, $\delta \sim 18°$, and $v_c \sim 5800$ km s^{-1}.

A few large voids were unambiguously identified and are visible in these cone diagrams. A void up to $v_c \sim 6000$ km s^{-1}, bounded by $3^h < \alpha < 4^h$, fills our declination range. At its lower velocity end, this void connects to a "local void" lying below ~ 4000 km s^{-1}at $3^h \lesssim \alpha \lesssim 8^h$; while at the high velocity end, it connects to another large void at $\delta \gtrsim 18°$ with a velocity stretching out to at least $10,000$ km s^{-1}. The existence of these voids clearly shows that the PP supercluster does not extend to $\alpha > 3^h$ within our survey declination range.

There is another giant void with $6000 \lesssim v_c \lesssim 8000$ km s^{-1}and $\alpha \lesssim 9^h$. This void is located just in front of the Great Wall and appears to extend well into the ZOA. As R.A. decreases, however, the galaxy density in this void increases while the contrast between the Great Wall and its neighboring region decreases.

5. Conclsuions

Our main results, based on the extinction-free galaxy distribution of a uniform sample of IRAS galaxies in the area of $2^h < \alpha < 10^h$ and $0° < \delta < 36°$, can be summarized as follows: (1) Our data extended some otically known concentrations and revealed some new ones, but the possibility of a nearby, very rich supercluster can probably be ruled out in this region. (2) Several voids are unambiguously identified. In particular, a large void between $\alpha \approx 3^h$ and 4^h, up to $v_h \sim 6000$ km/s, separates the Pisces-Perseus supercluster at $\alpha < 3^h$ from structures at $\alpha > 4^h$. (3) There are excessive galaxies around $v \sim 5000$ and ~ 8500 km s^{-1}, respectively. The latter "wall" appears to gradually diffuse out after it enters the Zone of Avoidance from the northern Galactic hemisphere.

Acknowledgments. We are grateful to the staffs of Arecibo Observatory for their help in our radio observations, and to W. Saunders and M. Strauss for providing is with their unpublished data. This work benefited from using the NED and was supported in part by a grant from NASA's IR, Submm and Radio Astronomy Program at JPL.

References

Böhm-Vitense, E. 1956, PASP, 68, 430

Burstein, D., & Heiles, C. 1982, AJ, 87, 1165

Dodd, R. J., & Brand, P. W. J. L. 1976, A&AS, 25, 519

Dow, M. W., Lu, N. Y., Houck, J. R., Salpeter, E. E., & Lewis, B. M. 1988, ApJ, 324, L51

Fitzgerald, M. P. 1974, A&A, 31, 467

Faber, S. M., & Burstein, D. 1988, in *Large Scale Motions in the Universe*, ed. V. C. Rubin & G. Coyne (Princeton: Princeton Univ. Press), 115

IRAS Point Source Catalog, Version 2. 1988, Joint *IRAS* Science Working Group (Washington, D.C.:GPO) (PSC)

Fisher, K. B. 1992, Ph.D. thesis, University of California, Berkeley

Focardi, P., Marano, B., & Vettolani, G. 1984, A&A, 136, 178

Geller, M. J., & Huchra, J. P. 1989, *Since*, 236, 897

Giovanelli, R., & Haynes, M. 1993, AJ, 105, 1271

Hayschildt, M. 1987, A&A, 184, 43

Haynes, M. P., & Giovanelli, R. 1988, in *Large Scale Motions in the Universe*, ed. V. C. Rubin & G. Coyne (Princeton: Princeton Univ. Press), 31

Kerr, F. J., & Henning, P. A. 1987, ApJ, 320, L99

Kraan-Korteweg, R. C. 1989, in *Workshop on Large-Scale Structures and Peculiar Motions in the Universe*, ASP Conf. Ser., eds. D.W. Latham & L.N. da Costa, Rio de Janeiro, 22-26 Map, 1989

Lu, N. Y., Dow, M. W., Houck, J. R., Salpeter, E. E., & Lewis, B. M. 1990, ApJ, 357, 388

Maurogordato, S., Proust, D., & Balkowski, C. 1991, A&A, 246, 39

Saunders, W., *et al.* 1991, Nature, 349, 32

Saitō, M., Ohtani, H., Asonuma, A., Kashikawa, N., Maki, T., Nishida, S., & Watanabe, T. 1990, PASJ, 42, 603

Saitō, M., Ohtani, H., Baba, A. Hotta, N., Kameno, S., Kurosu, S., Nakata, K., & Takata, T. 1991, PASJ, 43, 449

Seeberger, R., Huchtmeier, W. K., & Weinberger, R. 1994, A&A(preprint)

Strauss, M. A., Davis, M., Yahil, A., & Huchra, J. P. 1990, ApJ, 361, 49

Takata, T., Yamada, T., Saitō, M., Chamaraux, P., & Kazés, I. 1994, A&A (preprint)

Weinberger, R. 1980, A&AS, 40, 123

Yahil, A., Strauss, M. A., Davis, M., & Huchra, J. P. 1991, ApJ, 372, 380

Yamada, T., Takata, T., Djamaluddin, T., Tomita, A., Aoki, K., Takata, A., & Saitō, M. 1993, MNRAS, 262, 79

Discussion

D. Altschuler: What was your success rate for detections in HI?

N. Lu: We observed about 300 objects in HI, and detected about 80 galaxies. But we preferentially observed those sample sources not known to us (by then) to be galaxies, so this rate does not reflect the selection rate of galaxies by our colour criteria. In fact, about 85% of our sample sources above $|b| = 10°$ are now known to be galaxies.

H. van Woerden: Your intention is to use IRAS-selected samples to correct the incompleteness of the space distribution of optically detected galaxy samples. But is it not true that IRAS-selected samples also remain incomplete, since they miss certain types of galaxies (e.g. dwarfs)? If so, this underlines the importance of blind 21-cm surveys, and hence of receivers suitable for that purpose.

N. Lu: Yes, IRAS selected samples undersample certain types of galaxies. The purpose of this work is to generate a uniform sample of galaxies free from the Galactic extinction, and to compare it with the optically selected galaxy samples which are nonuniform due to Galactic obscuration.

W. Saunders: Your sample is 100 μm selected, but the histogram of 100 μm fluxes shows some evidence of incompleteness near the flux limit. As this incompleteness will correlate with 100 μm background, and this background correlates with extinction, surely there is some dependence of the probability of a galaxy entering your catalogue on optical extinction?

N. Lu: This might have to do with the PSC incompleteness near the Galactic plane. This could be tested by redoing the 100 μm flux plot separately for $|b| > 5°$ and for $|b| < 5°$.

Unveiling Large-Scale Structures Behind the Milky Way
ASP Conference Series, Vol. 67, 1994.
C. Balkowski and R. C. Kraan-Korteweg (eds.)

The quality of extragalactic IRAS selections

E.J.A. Meurs

ESO, Garching, Germany

Abstract.
IRAS sources that are likely to be extragalactic can be extracted from the Point Source Catalog with a high degree of reliability. Several projects have therefore selected such samples, for a variety of studies. It is of interest to quantify the success level of these selections. A method is described with which the successes and failures of sample selections can be assessed. Examples are given of application of such an assessment, and aspects particular to work in the Zone of Avoidance are emphasized. Finally it is attempted to intercompare the various extragalactic IRAS PSC selections that have been utilized.

1. Introduction

Extragalactic objects that occur in the IRAS Point Source Catalog (PSC) are clearly of interest to studies of galaxies behind the Galactic Plane because of the reduced effects of interstellar extinction as compared to optical wavelengths. In addition, the IRAS data cover almost all of the sky with a high level of homogeneity, and thus may allow following large-scale structures in the distribution of galaxies into and through the Zone of Avoidance.

Since the PSC has become available, several attempts have been made to extract the extragalactic objects from this database in a statistical fashion. The number of PSC sources strongly suggests such an approach: the entire catalogue contains 225,000 sources, the extragalactic objects are estimated to comprise some 30,000 of these (Soifer et al. 1987). Studies related to general characteristics of IRAS galaxies, such as overall sky distribution (Meurs and Harmon 1988) or dipole direction (Harmon et al. 1987), can be carried out after a statistical assessment of the success of the selection. More detailed projects, for instance modelling of the velocity field of the IRAS galaxies (Strauss et al. 1990) may wish to increase the quality of the association with extragalactic objects by follow-up observations but then are still helped by reliable candidates.

The attempts to retrieve the extragalactic objects from the PSC have mostly employed some combination of flux qualities and IR colours (based on the IRAS bands). The idea is normally to include the extragalactic PSC objects as completely as possible and to limit at the same time the contamination by different kinds of objects. In all these cases it is of interest to assess properly the quality of the selection. Whereas some authors have included checks of a particular aspect of their selection, mostly this is not practiced in a rigorous manner nor are the tests extended to the selection methods of other authors, for the purpose of comparing results. In this contribution a well-defined procedure for the assessment of IRAS (or, in fact, any other) selection of objects is advocated and applied to the main eight, or so, published IRAS galaxy selections. In this connection, the need for a few well-identified test fields in the sky is stressed which enable performance tests to be carried out. Before doing so, the features of one of the selections of IRAS galaxies are described, followed by an exposition of the one procedure that goes beyond the simpler, conventional methods and employs multivariate statistical methods.

2. IRAS galaxies and sky distribution

A pre-selection of IRAS PSC sources on the basis of flux qualities prevents that an excess of inappropriate sources has to be processed by a classifier. Various identification programmes have shown that stars generally are brighter at the shorter IRAS wavebands (12 and 25 μm), in contrast to galaxies (brighter at 60 and 100 μm). Correspondingly, stars tend to have reliable fluxes (flux qualities 1 or 2) at 12 and 25 μm and galaxies at 60 and 100 μm. Most of the selections of IRAS galaxies thus require reliable fluxes at 60 and/or 100 μm.

In one of the first attempts of this sort, Meurs and Harmon (1988) started with the 33435 PSC sources with reliable fluxes at 60 and 100 μm. After examination of the observed (PSC-) characteristics of a few classes of sources, isolated on the basis of their different sky distributions, two IRAS-colour criteria and one flux requirement (at 100 μm) were developed to discriminate between three (or four) categories of object. First, extragalactic objects. Second, a Galactic component strongly concentrated to the Galactic Plane (styled Thin Plane). Third, a Galactic component with broader distribution around the Galactic Plane (named Cirrus, for simplicity). (Fourth, stars).

The extragalactic component that was obtained in this way shows a quite homogeneous sky distribution, which does not suggest severe Galactic contamination. Several enhancements in the sky distribution of IRAS sources correspond evidently to galaxy clusters and chains known from optical data (including the Supergalactic Plane), albeit with lesser contrast due to the broad IR luminosity function of galaxies and the bias towards Spirals. A region along the central half of the Galactic Plane (between, roughly, longitudes +90 ° and -90 °) is on the other hand devoid of sources, as a result of source confusion. Meurs and Harmon (1988) made an attempt to describe quantitatively the efficiency of this selection, which indicated that over 90 % of the sources selected could indeed be galaxies, except in some Cirrus dominated regions (such as the Galactic Plane!). Similarly, some 90 % of all galaxies that are in the preselected sample (reliable fluxes at 60 and 100 μm) should be present in the extracted extragalactic component.

3. Multivariate selection of extragalactic PSC sources

A selection as just described can still be improved by means of more sophisticated statistical techniques. Proceeding as above, the source categories Extragalactic, Thin Plane and Cirrus were again sampled on the basis of their respective sky distributions and considered as so-called training sets (this time the stars were at forehand essentially cut out by a simple 25 to 60 μm colour criterium). Each of these source categories is represented by a clustering of data points in a four-dimensional feature space: three IR colours and one flux (at 100 μm again). These data point distributions are interpreted as the probability distributions to find a source of a particular category at a certain position in this n-dimensional feature space. Equal probability between the source categories defines decision surfaces (here 3-dimensional hypersurfaces) that partition the feature space into regions where one of the categories has the greatest likelihood. Each data point falls into one of these cells and can thus be attributed to one of the classes of object. The result is a Maximum Likelihood classification of the complete set of data points.

Meurs, Adorf and Harmon (1987) show the three IRAS PSC components selected in this way. The extragalactic sources exhibit a sky distribution quite similar to the one obtained in the previous section. The advantages of this technique are the repeatability and potential objectiveness of the procedure. Furthermore, the class-specific cells of the n-dimensional feature space are established in a more flexible way that takes into account the actual shape of the relevant distributions of data points and that allows more appropriately shaped divisions between the various categories.

4.   Quantitative assessment of extragalactic IRAS selections

4.1.   Figures of merit for selected samples

As for the selection in Section 2, it is important to judge the results of a classifier by means of objective assessment procedures. Prusti et al. (1992, see their Appendix) have endeavoured to develop a general scheme of quantitative evaluation of classification results. It is appropriate then to make distinction between classifier performance and selection result. The performance of a classifier can conveniently be described by an Operating Characteristic (e.g. Melsa and Cohn 1978) which shows the interdependence of hit rate (completeness) and false alarm rate (potential contamination) and allows a deliberate choice for the Operating Point of the classifier. Examples of this are in Prusti et al. for selections of Young Stellar Objects from the IRAS PSC. For the samples that are selected it appears more useful to estimate, besides completeness, the actual contamination level in the resulting set of objects (see Prusti et al. for further details). This can also be expressed as a success rate, which equals { 1 – contamination level }.

4.2.   Selections considered

The two figures of merit, completeness and contamination level, will now be investigated for several published selections of extragalactic IRAS sources. The authors are listed in Table 1, which should include the main current selections that are around. In a few cases authors have themselves discarded previous versions of their selections, which then have been left out from the table.

All the groups in the table employ a flux quality selection. Except Meurs, Adorf and Harmon (1988), all other selections rely on colour– and/or flux criteria applied to the IRAS PSC data. It should be emphasized that the various groups

Table 1.   Selectioneers of extragalactic IRAS sources

Authors	Year	Abbreviation
Meurs,Harmon	1988	MH
Meurs,Adorf,Harmon	1988	MAH
Strauss et al.	1990	SD
Yamada et al.	1993	Yam
Babul,Postman	1990	BP
Lu et al.	1990	Lu
Yahil,Walker,Rowan-Robinson	1986	YW
Rowan-Robinson	1988	RR

may have had different aims in mind, which makes straightforward comparisons not completely fair. For instance, Babel and Postman (1990) focussed on galaxy selections at comparatively high Galactic latitudes, while Lu et al. (1990) in contrast wanted to find galaxies in Cirrus contaminated regions. Strauss et al. (1990) on the other hand needed galaxy candidates all over the sky and were willing to accept substantial contamination for trying to reach high completeness, knowing that they would inspect all selected candidates in a programme of optical follow-up observations; they stopped at a relatively high flux density limit in order to keep this task manageable. Meurs, Adorf and Harmon simply wanted a map of the extragalactic IRAS sky, as homogeneous as possible over all regions of the sky, and down to low flux levels.

For the sake of the comparison in the next subsection, the question that is posed is just: how well do the various selections in selecting galaxies in regions with high contamination, such as the Galactic Plane, irrespective of any particular flux limit.

4.3.   Results for two test fields

In order to judge the effectiveness of their classifier, Prusti et al. (1992) examined the selection results in a region with many confirmed Young Stellar Objects; other sources in this area (around the Chamaeleon I star forming region) were classified from catalogued information or directly from optical sky survey plates.

For the extragalactic IRAS sources one would like again to use regions with identifications for the IRAS PSC sources as complete as possible. The best field for this is the South Galactic Pole where Wolstencroft et al. (1986) obtained almost complete identifications. However, the potential contamination by Cirrus-like Galactic sources is close to zero and any evaluation of extragalactic IRAS selections is meaningful for general selection effectiveness but less relevant to studies in the Zone of Avoidance. A sub-area of this field is used containing hardly any unidentified source.

Unfortunately near the Galactic Plane no such field is available at present. As a reasonable approximation therefore, the Chamaeleon I region may serve as an example of a region where galaxies have to be found amidst contaminating Galactic sources. The area used here is chosen somewhat smaller than in Prusti et al.; catalogued or published information has again been used and been complemented by visual inspection of sky survey plates. The total number of galaxies in this test field is now on the other hand a bit small; further, one has to keep in mind that the contaminators are mainly Cirrus, not Thin Plane sources (the field is at b ∼ 16 °).

The prescriptions of each of the selectioneers listed in Table 1 have been followed to reproduce their selection in each of the two fiels, South Galactic Pole and Chamaeleon I. The former informs mostly about completeness, the latter shows mainly the susceptibility for contaminators. The derived numbers for completeness and contamination are presented in Table 2 for SGP and Table 3 for Chal. The results are given both as fractions (which show the actual numbers of object that are involved) and as percentages. For the completeness level two values are given, the first applies to the total PSC, the second one (between parentheses) refers to the preselection by flux qualities which is, in a way, informative about the effectiveness of the classifier as such.

Table 2.   South Galactic Pole field

Authors	Completeness (total PSC)	Completeness (preselected)	Contamination
MH	32/48=67%	(32/37=86%)	0 %
MAH	29/48=60%	(29/37=78%)	0 %
SD	44/48=92%	(44/45=98%)	0 %
Yam	43/48=90%	(43/44=98%)	0 %
BP	30/48=63%	(30/39=77%)	0 %
Lu	20/48=42%	(20/40=50%)	0 %
YW	39/48=81%	(39/39=100%)	0 %
RR	45/48=94%	(45/45=100%)	0 %

Table 3.   Chamaeleon I field

Authors	Completeness (total PSC)	Completeness (preselected)	Contamination
MH	2/6=33%	(2/3=67%)	2/4=50%
MAH	2/6=33%	(2/3=67%)	0/2=00%
SD	6/6=100%	(6/6=100%)	19/25=76%
Yam	5/6=83%	(5/6=83%)	8/13=62%
BP	5/6=83%	(5/6=83%)	10/15=67%
Lu	3/6=50%	(3/3=100%)	1/4=25%
YW	6/6=100%	(6/6=100%)	22/28=79%
RR	5/6=83%	(5/6=83%)	9/14=64%

5. Conclusion

The numbers in Tables 2 and 3 allow a comparison of the various selections that were published. Due to the different characteristics of the two test fields SGP and Chal (see above), the contamination in the SGP region remains at zero level. The values for contamination obtained in Chal are more meaningful and of greater relevance to the Zone of Avoidance. Inspection of the numbers in these tables shows how well each of the selections performs in regions with serious contamination. Only a few noticeable points are mentioned here. The survey of SD aims at reaching high completeness, which is achieved at the cost of increasing contamination; this is clear from the numbers in the tables. The substantial contamination is accepted since an extensive programme of optical follow-up observations is carried out. A similar statement can be made about RR. BP wanted an extragalactic IRAS sample only at high Galactic latitudes, but their completeness in the SGP field is not very favourable. Lu has the lowest completeness in the SGP field, but his specific intention was a selection in contaminated regions; his completeness level in ChaI shows that his criteria work quite well here. MH and MAH have the lowest completeness in the presence of strong contamination, but for MAH the contamination level remains at zero even in ChaI, probably as a result of employing the most advanced selection method.

As alluded to above, the status of the two test fields that could be used for this pilot comparison and the different intentions of the various selectioneers render firmer conclusions premature. The authors emphasized different environments in the sky and often stopped at comparatively high flux limits. Therefore, as indicated before, the basic question considered here is simply how well each of the selections in Table 1 would do if applied at low Galactic latitudes. The ChaI field needs anyway to be extended in order to include more extragalactic IRAS sources that in principle could be selected. At the moment the numbers in Table 1 clearly suffer from small number statistics. Identifications in a suitable test field in the Galactic Plane are recommended for the near future.

References

Babul, A., and Postman, M. 1990, ApJ, 359, 280

Harmon, R.T., Lahav, O., and Meurs, E.J.A. 1987, MNRAS, 228, 5P

Lu, N.Y., Dow, M.W., Houck, J.R., et al. 1990, ApJ, 357, 388

Melsa, J.L., and Cohn, D.L. 1978, Decision and Estimation Theory, McGraw-Hill, New York

Meurs, E.J.A., Adorf, H.-M., and Harmon, R.T. 1988, in Astronomy from Large Databases – Scientific Objectives and Methodological Approaches, A. Heck and F. Murtagh, ESO, Garching, 49

Meurs, E.J.A., and Harmon, R.T. 1988, A&A, 206, 53

Prusti, T., Adorf, H.-M., and Meurs, E.J.A. 1992, A&A, 261, 685

Rowan-Robinson, M. 1988, in Comets to Cosmology, A. Lawrence, Springer-Verlag, Heidelberg, 348

Soifer, B.T., Houck, J.R., and Neugebauer, G. 1987, ARA&A, 25, 187

Strauss, M.A., Davis, M., Yahil, A., and Huchra, J.P. 1990, ApJ, 361, 49

Wolstencroft, R.D., Savage, A., Clowes, R.G., et al. 1986, MNRAS, 223, 279

Yahil, A., Walker, D., and Rowan-Robinson, M. 1986, ApJ, 301, L1

Yamada, T., Takata, T., Djamaluddin, T., et al. 1993, ApJS, 89, 57

Discussion

N. Lu: Are the statistical numbers you showed on the selection method of Lu *et al.* (1990) based on their original or improved criteria?

E. Meurs: They were based on the improved criteria, thank you for pointing this out.

O. Lahav: What is the algorithm of your classifier? In particular, is it non-parametric? Does it allow curved boundaries? And does it yield a Bayesian classification? There are methods (e.g. Artificial Neural Network) which allow these properties.

E. Meurs: The classifier you are referring to, the supervised classification, is a Maximum Likelihood result where the distribution of points in parameter space is only approximated by n-dimensional Gaussians. We have been considering, and been making some first attempts, to utilize other, more adaptable methods like near-neighbours analysis and neural networks.

Unveiling Large-Scale Structures Behind the Milky Way
ASP Conference Series, Vol. 67, 1994.
C. Balkowski and R. C. Kraan-Korteweg (eds.)

QUANTIFYING LARGE SCALE STRUCTURE BEHIND THE MILKY WAY USING THE IRAS POINT SOURCE CATALOG

W.SAUNDERS, W.J.SUTHERLAND, G.EFSTATHIOU, H.TADROS
Astrophysics, NAPL, Keble Road, Oxford OX1 3RJ, UK

S.MADDOX
RGO, Madingley Road, Cambridge CB3 0EZ, UK

R.G.MCMAHON, S.D.M.WHITE
Institute of Astronomy, Madingley Road, Cambridge CB3 0HA, UK

M.ROWAN-ROBINSON, S.J.OLIVER, O.KEEBLE
Blackett Laboratory, Imperial College, Prince Consort Road, London SW7 2BZ, UK

C.S.FRENK
Department of Physics, South Road, Durham DH1 3LE, UK

J.V.SMOKER
NRAL, Jodrell Bank, Macclesfield, Cheshire SK11 9DL, UK

ABSTRACT

1. A progress report is given on a redshift survey of IRAS galaxies selected from the Point Source Catalog   (the PSC-z survey). The final catalogue will consist of 15,000 galaxies with $S_{60} \geq 0.6$ Jy over 84% of the sky, and will extend right through the Plane at the anti-centre. We expect to complete the survey in 1994.

2. Selection criteria are presented for galaxies in the PSC which allow uniform coverage over 92% of the sky, with 90% galaxy completeness everywhere. At low latitudes, external galaxies selected by these criteria are outnumbered by sources in our own galaxy; we discuss techniques for weeding out these local sources.

INTRODUCTION

The IRAS survey has transformed our understanding of the dynamics of the Local Universe by allowing the compilation of uniform, almost all-sky redshift surveys of galaxies that trace the mass distribution very well (Kaiser *et al.* 1991, Dekel *et al.* 1994). Following the success of the QDOT 1-in-6 survey, we embarked on a redshift survey of the remaining IRAS PSC galaxies without known redshift, in order to create a sample with (a) a higher sampling rate locally, and (b) a much larger effective volume, for cosmological study. The purpose of this new 'PSC-z survey' is to understand the origin of the local velocity field, to test whether light traces mass on large scales, and to test whether the statistics of the galaxy distribution are compatible with structure formation from inflation via gravitaional instability.

SAMPLE SELECTION

The starting point for our survey is the QMW IRAS Galaxy Catalogue (QIGC) (Rowan-Robinson *et al.* 1990, henceforth RR90) from which the QDOT survey was selected (Lawrence *et al.* 1994). We have however supplemented the QIGC in various ways to further improve its sky coverage and completeness; we are also treating extended sources slightly differently.

Areas excluded from the PSC-z survey are:

(i) the IRAS coverage gaps which failed to get 2 Hours-Confirmed passes.

(ii) Areas which are flagged as High Source Density at 12 or $25\mu m$ (where stellar source density makes identification of galaxies impossible) or at $60\mu m$ (where differences in the PSC extraction software sacrificed completeness for the sake of reliablity).

(iii) areas where $I_{100} > 25\,MJy/ster$ (as defined by Rowan-Robinson *et al.* 1991, and roughly corresponding to CIRRUS3=50); above this level, extinctions are typically greater than $A_V = 1.5^m$, making optical identifications and spectroscopy uncertain, and also the real galaxies are outnumbered by local IRAS sources with galaxy-like far-infrared colours. In estimating the extinction from I_{100}, point sources have been subtracted and = a small adjustment has been made to take into account large scale dust temperature variations in the Galaxy.

(iv) areas within 2.5° and 5° of the SMC or LMC with $I_{100} > 5$ or $10\,MJy/ster$ respectively are also excluded.

The mask includes the areas suffering from noise lagging (Beichmann *et al.* 1988, henceforth PSC-ES), and hysteresis has been shown to be a very weak effect by Strauss *et al.* 1990. Our overall sky coverage is 84% of the sky. The resulting mask and basic catalogue are shown in Figure I; our criteria allow us to go right through the Galactic Plane in some parts of the anti-centre.

Because we have access to optical information and intend to get redshifts for all galaxy sources, the very high reliability of the QIGC (in the sense that 98%

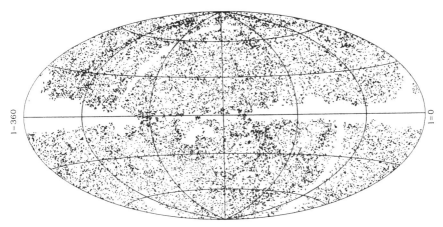

1=360

l=0

FIGURE I   Sky distribution in Galactic coordinates of the 2-D parent
sample and the mask used in constructing the PSC-z survey. Some contamination of the sample by local sources with galaxy-like colours is apparent
at low latitudes.

Thus we have been able to relax the colour and identificational criteria used
in the QIGC, in order to pick up the small number of galaxies excluded from
QIGC, and we have been able to extend the survey into regions which are High
Source Density at 100μm only. The cost of this is significant contamination of
the 2D catalogue at low latitude by local sources (Figure I).

2HCON areas

In 2 HCON areas, where the PSC completeness is not guaranteed to 0.6 Jy, we
have supplemented the catalogue with 1HCON sources with galaxy-like colours
from the Point Source Catalog   Reject File, where there is a corresponding
source in the Faint Source Survey. This has revealed many sources where two
individual HCON detections failed to be merged into a PSC source (PSC-ES
XII.A.3), as well as sources which failed at least one HCON for whatever reason.
The completeness in these areas has been increased from 95% (PSC-ES XII.A.4)
to at least 98%.

Extended Sources

All IRAS surveys are bedevilled by the question of how to deal with galaxies
which are extended with respect to the IRAS 60μm beam (approx $1' \times 4'$).
The approach we have settled on is to preferentially use PSC fluxes, except for
sources badly confused, or identified with galaxies large enough to have their
flux significantly underestimated ($S_{60,PSC}/S_{60,true} < 0.9$) in the PSC. We are
assembling a new compilation of the UGC/ESO/ESGC (Corwin, priv.com.) in
order to select all galaxies in the area of our survey with extinction-corrected
diameter $d_{25} > 3'$, using the corrections of Cameron 1990. The diameter scales

 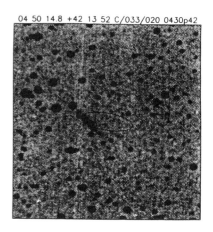

FIGURE II   4' x 4' APM raster scans of POSS plates at (a) moderate ($(l, b) = (150, 20)$) and (b) low ($(l, b) = (163, 1)$)) latitude, showing the IRAS error ellipses. Intensity has been logarithmically scaled to emphasis low surface brightness features.

of the optical catalogues will be matched by demanding a uniform diameter/FIR relationship for the galaxies in each. All such large galaxies will be addscanned and added to the catalogue if $S_{60,addscan} \geq 0.6$ Jy. This procedure has the benefit of uniformity across the sky and ensures completeness for extended galaxies.

OPTICAL IDENTIFICATIONS

For all sources, digitised optical material has been obtained from APM or COS-MOS scans of POSS or SRC sky survey plates. This gives magnitudes and arcsecond positions of all feasible candidates. The actual identifications are made using the likelihood analysis of Sutherland and Saunders (1992) and this gives secure identifications in virtually all cases. Examples of APM raster scans of a POSS plate at low latitudes are shown in Figure II.

At this stage, local sources are rejected by a combination of their IRAS colours, appearence on sky survey plates and information from the SIMBAD database. Of the 17,000 sources in the sample, 15,250 are confirmed galaxies, 1500 are local sources, and 250 have no clear identification although their IRAS properties are consistent with them being galaxies.

REDSHIFT AQUISITION

We have undertaken a very large literature search for IRAS galaxy redshifts. At the outset of the survey (1991), about two-thirds of the galaxies in our catalogue had known (but often unpublished) redshifts from a very wide range of sources.

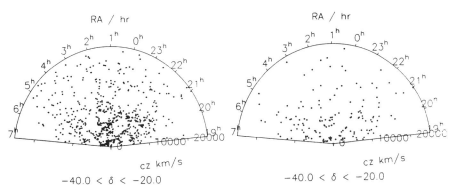

FIGURE III   (a) Cone plot of a completed part of the survey, representing about one tenth of our overall sky coverage. (b) the equivalent QDOT figure.

It is this that made the survey feasible in realistic amounts of telescope time. We were fortunate enough to be awarded time on the INT, AAT and CTIO 1.5m to measure redshifts for the remaining galaxies. We have taken about 4000 redshifts ourselves to date, with 500 still required as of January 1994.

We use low dispersion Faint Object Spectrographs because of their very high efficiency and low overheads. For the emission line galaxies in our survey, adequate spectra are usually obtained in 100-200s, allowing 100 redshifts per night at high latitude and 75-80 in the Plane. To squeeze the maximum accuracy out of our spectra, we model the optimally extracted and continuum subtracted 1D spectrum as an arbitrary blend of $H\alpha, NII, SII$, and estimate the redshift and internal error by least χ^2 fitting. Internal errors are typically better than $100\,\mathrm{km\,s^{-1}}$ while slit centering errors are expected to be of the same order. Wavelength calibration is based on cross-correlation with a well-calibrated sky spectrum, the error from this is small. The target is $100 - 150\,\mathrm{km\,s^{-1}}$ total external error, versus $250\,\mathrm{km\,s^{-1}}$ for QDOT.

A cone plot for a completed slice of the survey, and the comparable plot for the QDOT survey, is shown in Figure III. We expect to complete redshift aquisition during the winter of 1994-95.

ACKNOWLEDGEMENTS

This survey has only been made possible by the generous assistance from many people in the astronomical community. We are particularly grateful to John Huchra, Tony Fairall, Karl Fisher, Michael Strauss, Marc Davis, Raj Visvanathan, Luis DaCosta, Riccardo Giovanelli, Nanyao Lu, Carmen Pantoja, Tadafumi Takata, Tin Conrow, Mike Hawkins, Delphine Hardin, Mick Bridgeland, Renee Kraan-Kortweg, and the staff at the INT, AAT and CTIO telescopes.

EXTENSION TO LOWER LATITUDES

The area of the main PSC-z survey is determined by our ability to make optical identifications and obtain redshifts. Sky coverage could be increased by extending the survey into areas with $I_{100} > 25\,\mathrm{MJy/ster}$. The source counts for IRAS sources with galaxy-like colours in these low-latitude regions (Figure IV) show that (a) for areas with $I_{100} < 100\,\mathrm{MJy/ster}$, there is no turnover in the counts at low fluxes until $0.75\,\mathrm{Jy}$, suggesting that incompleteness is not a serious problem to this flux level, and (b) there is contamination by local sources in our own Galaxy, especially at high fluxes. These local sources consist of:

(i) Sources associated with pre- or post-main sequence stars with warm IRAS colours; bipolar nebulae, T Tauri stars, Ae/Be stars, planetary nebulae.
(ii) Reflection nebulae around bright stars, with galaxy-like IRAS colours at $25, 60, 100\mu m$ but often a warm $12/25$ colour from the Rayleigh-Jeans tail.
(iii) HII regions, often warm but sometimes indistinguishable from galaxies.
(iv) Cold sources associated with molecular clouds or infrared cirrus.

The criteria we have devised to maximise the number of galaxies included, while minimising the level of contamination by these local sources, is as follows:
(i) Exclude all sources with $S_{60} > S_{100}$, or with $25\mu m$ detections and $S_{25} > 0.5S_{60}$. These two conditions exclude almost all stars and planetary nebulae, at the cost of also excluding Seyfert galaxies (about 6% of IRAS galaxies at high latitudes, defined as areas where $I_{100} < 10\,\mathrm{MJy/ster}$).
(ii) Exclude all sources with $12\mu m$ detections and $S_{12} > 0.25S_{60}$. This efficiently excludes remaining bright stars, at a cost of excluding just 22 galaxies at high latitiudes (0.2%, of which half are already excluded by (i)).
(iii) Exclude all sources with $100\mu m$ detections and $S_{100} > 5S_{60}$. This excludes myriads of cirrus sources, and the condition is violated by only 20 known galaxies at high latitudes (0.2%). Unfortunately, these are all large nearby galaxies, just the sort we would like to find behind the Plane.
(iv) Exclude all sources with Correlation Coefficient at $60\mu m < 0.98$ (i.e not A,B,C). This very efficiently excludes cirrus sources, while only 82 galaxies are excluded at high latitudes (0.8%). Again, these are preferentially nearby.

Figure V shows the $25/60$ and $60/100$ ratios, and $CC(60)$ for all galaxies in our catalogue at high galactic latitude within $5,000\,\mathrm{km\,s^{-1}}$. The symbol size indicates nearness; note that even the nearest galaxies usually have $CC(60){=}A,B,C$. Overall, the above conditions exclude less than 10% of high latitude galaxies, although this figure rises to 20% for the very nearest ($V < 1000\,\mathrm{km\,s^{-1}}$) galaxies. There does not seem any easy way of keeping more of these nearby galaxies without including huge numbers of cirrus sources also. However, these extended, cool, gas rich, nearby galaxies should be easily detected in the blind HI searches already discussed by Butler and Stewart in these proceedings, so IRAS and HI searches should be seen as complementary.

We have not attempted to use $100\mu m$ upper limits in our criteria, because these are so often corrupted in the Plane as to be unuseable. Even our condition (iii), while much milder than than of e.g. Yamada 1993 or RR90, will lead to a few real galaxies being excluded due to corrupted $100\mu m$ flux. Nor have we used the

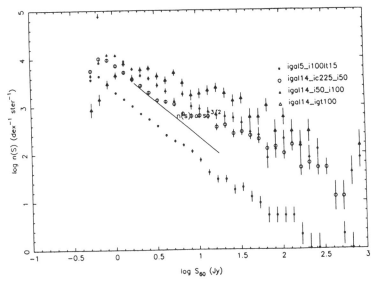

FIGURE IV   Source counts for sources with galaxy-like colours at low latitudes with I_{100} in the range 25-50 MJy/ster (open circles), 50 – 100 MJy/ster (filled triangles) and > 100 MJy/ster (open triangles) vs galaxy counts at high latitudes (filled circles).

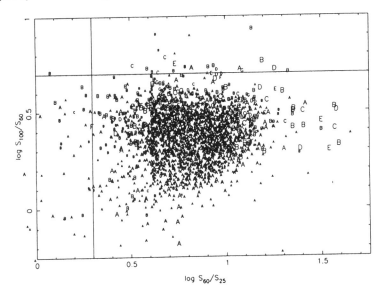

FIGURE V   25/60μm and 60/100μm colours for IRAS galaxies with $S_{60} \geq 1$ Jy and $V < 5000$ km s^{-1} at high latitudes, with the symbol showing the Correlation Coefficient (A,B,C etc). Symbol size is scaled as inverse square root of distance. The proposed colour cuts are shown by horizontal and vertical lines.

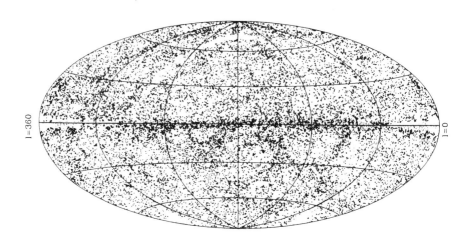

FIGURE VI   Maximum sky coverage for uniform IRAS PSC-selected samples. The mask consists of the coverage gaps, HSD areas at $12, 25$ or 60μm, and areas with $I_{100} > 100\,$MJy/ster. The total sky coverage is 92%.

SES2 flag, on the basis that (a) many associations with small extended sources will be serendipitous, and that (b) we would not wish to evict large nearby galaxies if we can help it.

The sources selected by these criteria are shown in Figure VI. Also shown is the residual mask, consisting of the coverage gaps, areas which are HSD at $12, 25$ or 60μm, and areas with $I_{100} > 100\,$MJy/ster. The total sky coverage is 92%.

This procedure allows a further 1.1 steradians of sky to be used, and gives a further 2000 sources with $S_{60} \geq 1\,$Jy at 60μm. 250 of these are confirmed galaxies with measured redshifts; based on the galaxy source counts at higher latitudes and completeness discussed above, we estimate that one third of the total are external galaxies, i.e. a further 400 remain. Although this may seem a modest success rate, it is in line with the detection rates found by Lu *et al.* (1990) and Bottinelli *et al.* (Priv. Com.) for their HI integrations on IRAS sources at low latitudes, and in line with the contamination expected given the source counts shown in Figure IV. At present, it is the only realistic way of uniformly selecting galaxies at very low latitudes.

Other sets of colour criteria have been proposed, with comparable and sometimes better efficacy (Yamada 1993, Meurs 1994, Rowan-Robinson 1990, RR90, Lu *et al.* 1990). However, the primary emphasis here is on *uniformity*, so that the resulting samples can be used for quantitative analysis of Large Scale Structure, with reasonable completeness even at low redshifts a second priority.

Reducing the Contamination

Before HI or optical followup is attempted, it is clearly essential to reduce the level of contamination by local sources. We have explored the following ways of doing this:

(i) The SIMBAD database has allowed us to exclude hundreds of sources as confirmed PN's, HII regions, Young Stellar Objects, Molecular Clouds or other sources in our own galaxy.

(ii) Examination of sky survey plates allows the rejection of many further sources as obvious dark clouds, reflection nebulae or bright stars. Many galaxies are also identifiable from the plates. The major ambiguity is between reflection nebulae and galaxies; usually these can be resolved by comparing plates of two different colours, e.g. blue and I. We have not found any one colour generally superior in either the POSS or SRC surveys; the bluer plates suffer higher extinction but are generally deeper.

(iii) Examination of the raw or coadded IRAS data, for example coming out of the addscan procedure, can often allow cirrus sources to be identified immediately, by the high signal to noise structured and extended emission seen at both 60 and 100μm.

(iv) CO spectroscopy:
A powerful discriminant is provided by the CO measurements such as those of Wouterlout and Brand (1989). Sources associated with molecular clouds have copious CO emission at very low velocity and narrow velocity width. Sources not detected in CO, or detected at a flux and/or linewidth not satisfying the expected FIR-CO relationship, are expected to be either galaxies, bright Vega-excess-type stars, HII regions, or planetary nebulae. The latter three types are easily excluded by their appearance on sky survey plates for all but the most extreme levels of extinction.

(v) VLA snapshots:
The tight FIR/radio correlation (e.g. Unger *et al.* 1989) in principal offers a method for obtaining higher resolution maps of obscured IRAS galaxies. We have obtained a total of 14 hours of VLA snapshots at 20cm as a pilot project for this idea, demonstrating that the approach is feasible, with some good galaxy candidates being found even in regions of high obscuration. However, the similarity of the FIR/radio relationship for galaxies, planetary nebulae and HII regions (e.g. Zijlstra *et al.* 1988, Meinert *et al.* 1990), the comparatively long integration times needed even at the VLA (6-12 sources per hour), and the significant fraction of confused fields limit the viability of this approach. At 6cm the confusion problem is reduced but integrations are longer. The VLA 20cm All-Sky Survey will however, be very useful in allowing sources brighter than 1 Jy but without radio counterparts to be rejected.

(vi) NIR imaging:
The most promising approach is K'-band imaging of IRAS galaxy candidates. NICMOS3 detectors on 2m class telescopes are able to image the entire IRAS error ellipse in a single integration, and integration times of 60s will pick up the overwhelming majority of IRAS galaxies. Stellar densities are $10^4 - 10^5 \, deg^{-2}$ (Mamon, Cohen, priv. com.), i.e. very high but well below the confusion limit for images taken in reasonable seeing and with pixel sizes $\lesssim 1''$. We are undertaking a program to systematically image all galaxy candidates with $S_{60} \geq 1 \, Jy$.

Spectroscopy

Redshifts for sources with an optical identification can be obtained in 5-10 mins on 2m telescopes. For other sources, 21cm spectroscopy is required. The survey by Lu *et al.* 1990 showed that HI redshifts could be routinely obtained for completely obscured IRAS galaxies in the Plane, and many large radio telescopes are now able to simultaneously search a wide redshift range of e.g. $0-10,000\,\mathrm{km\,s^{-1}}$. However, integration periods can very long, e.g. typically several hours on the 76m Lovell Telescope for reasonable completeness to $10,000\,\mathrm{km\,s^{-1}}$. We hope to obtain up to six weeks time later this year on the Lovell Telescope for spectroscopy of sources already determined to be probable galaxies by methods (i)-(v) above.

COORDINATION

There are many groups pursuing HI or optical spectroscopy of IRAS or optically selected samples in the Plane, and coordination of these efforts avoids pointless duplication of effort. We maintain a comprehensive database of information on galaxies in the PSC; who has a redshift (published or not); who is planning to take a redshift; identification as non-galaxy source etc. We are happy to pass this information on for any desired portion of sky, in return for updated information. Contact w.saunders@physics.ox.ac.uk for further details.

REFERENCES

Beichmann *et al.* 1988. IRAS Explanatory Supplement.
Cameron, L.,M. , 1990. *Astron. Astrophys.* **233**, 16.
Dekel *et al.* 1994. *Astrophys. J.* **412** 21.
Kaiser et al. 1991. *Mon.Not.R.astr.Soc.* **252** 1.
Lawrence *et al.* 1994, in preparation.
Lu *et al.* 1990. *Astrophys. J.* **357**, 388.
Meinert *et al.* 1990. *Ast. Ges. Abstr. Ser* 5 20. Meurs 1994. These proceedings.
Rowan-Robinson *et al.* 1990. *Mon.Not.R.astr.Soc.* **253** 485.
Rowan-Robinson 1990. In *Comets to Cosmology*, Springer-Verlag.
Rowan-Robinson *et al.* 1991. *Mon.Not.R.astr.Soc.* **249** 729.
Strauss *et al.* 1990. *Astrophys. J.* **361** 49.
Sutherland and Saunders 1992. *Mon.Not.R.astr.Soc.* **259** 413.
Wouterlout and Brand 1989. *Ast. Ast. Supp.* **80** 149.
Unger *et al.* 1989. *Mon.Not.R.astr.Soc.* **236** 447.
Yamada *et al.* 1993. *Astrophys. J.* **89** 57.
Zijlstra *et al.* 1988. *Ast. Ast. Supp.* **79** 329.

Discussion

E. Meurs: You refer perhaps to use of the addscan information, which shows differences dependent on object type. How does this compare with the correlation coefficients?

W. Saunders: The correlation coefficients are a measure of both signal-to-noise and extension. Even clearly extended, nearby FIR-bright objects usually have good correlation coefficients. So the addscans really do contain more information than the PSC flags.

F. Kerr: You spoke of HI redshifts taking up to a few hours, you should be able to get them in shorter time than this.

W. Saunders: The figure I quoted was based on an estimate from Jodrell Bank for the MK1A 76m, for a flux of a few Jy km s^{-1} at say 10,000 km s^{-1}. Of course, most of the integration time will be spent on the hardest sources and it may well be worth shortening the integration times at the expense of completeness. I would be very pleased if my estimates are too pessimistic!

T. Yamada: Will pointed out the danger of using 100 μm data in infrared criteria. I agree with him. In our search we used the f_{60}/f_{100} ratio as selection criterion. The results have been compared with that of Strauss *et al.*'s survey (1.2 Jy sample), in which they did not use 100 μm data. We could detect about 95% of Strauss's galaxy at $\ell \lesssim 315°$, but we miss $\sim$ 30% at $\ell \gtrsim 315°$.

Search for IRAS Galaxies behind the Milky Way and the Puppis Hidden Concentration of Galaxies

T. Yamada

Department of Astronomy, Faculty of Science
Kyoto University, Sakyo-ku, Kyoto, 606-01, Japan

Abstract. We have systematically investigated the distribution of *IRAS* galaxies behind the Milky Way. Using this sample of *IRAS* galaxies, we found that some large-scale firamentally structures of galaxies are crossing the Milky Way region on the sky. We also present new radial velocity measurements of glaxies in the Puppis region and show there does exist a large concentration of galaxies at $cz \sim 2000$ km s^{-1}, which may have significant contribution to the Local Group motion.

1. Introduction

The region of the sky behind the Milky Way has long been a "zone of avoidance" (ZOA) of searching for galaxies. The large Galactic extinction prevents us from systematically investigating the distribution of galaxies in this region. The Milky Way region ($| b | < 15°$), however, covers a quarter of the whole sky in solid angle, and it is thus generally important to reveal the galaxy distributions in this region in order to understand overall galaxy distributions and motions in the Local Universe ($cz \lesssim 10,000$ km s^{-1}). Our main goals of stdying galaxy distribution behind the ZOA are (1) recognizing large-scale structures of galaxies crossing the Milky Way on the sky, and (2) finding previously unknown galaxy concentration behind the Milky Way which may have significant effects on the Local Group motion. Here we present the results of our systematic investigation of *IRAS* galaxies behind the Milky Way and report the new redshift measurements of galaxies in the Puppis region where a nearby galaxy concentration has been discovered just behind the Galactic Plane.

The contents of this paper and related works have already published in the following papers. The procedures and results of the *IRAS* galaxy search at southern Milky Way region are described in Yamada et al. (1993b). The existence of large-scale coherent structures crossing the Milky Way region is shown in Yamada et al. (1993a). The radial velocity distribution of galaxies in the Puppis "hidden" concentration is studied in Yamada et al. (1994). The effect of the Puppis cluster on the peculiar velocity of the Local Group is discussed in Lahav et al. (1993). Results of a search and redshift survey of *IRAS* galaxies behind the northern Milky Way region between $150° < l < 240°$ is shwon in Takata et al. (1994). Apparent magnitudes of galaxies detected behind the Milky Way is discussed in Saitō, Takata, and Yamada (1994). Revealing more larger-scale galaxy distribution towards $-$SGZ direction is another interesting subjects that we have been working on. The Monoceros supercluster at $(l, b) \sim (225°, 5°)$ and

$\sim 120\ h^{-1}$Mpc and the foreground low density regions are studied in Yamada and Saitō (1993).

2. Search For IRAS Galaxies behind the Milky Way

In this section, we briefly describe our systematical search for $IRAS$ galaxies behind the almost whole southern Milky Way region between $l = 210°$ and $360°$ at $\mid$ b $\mid< 15°$.

There are some advantages of using a sample of $IRAS$ galaxies rather than optically selected one. Because the far-infrared flux at 60μm little suffers from Galactic extinction, the $IRAS$ galaxies identified in the Milky Way region still form a lower limit sample of a flux limited sample; i.e., when we find a region where number density is higher than the average, we can recognize it as a real density-enhanced region even if the sample is not truely complete. On the other hand, it is very difficult to define an optical diameter or magnitude limited sample at low Galactic latitudes because Galactic extinction varies from place to place. Moreover, $IRAS$ sources have well defined positions, which makes our search rather easy. In order to reveal the overall distribution of galaxies behind the Milky Way, it thus might be even better to use $IRAS$ galaxies.

At the lower Galactic latitudes, $IRAS$ $Point$ $Sources$ $Catalog$ (IPSC) is dominated by a huge number of Galactic objects and thus we need some identification procedure to select out only galaxies from IPSC. We performed the search in two steps. Firstly, we selected $IRAS$ galaxy candidates from IPSC applying infrared color and flux-density criteria as previous authors did $outside$ the zone of avoidance. In the second step, we visually searched for the objects with galaxy image on the UK Schdmit Infrared (I-band) and IIIa-J Atlases. Though the Galactic extinction prevents more or less such optical identifications, we could detect such a large number of $IRAS$ galaxies that we can use them as a lower limit sample in comparison of their density with the whole-sky average. We define our infrared criteria as summarized in Table 1. To exclude stars, we use

Meaning of Criterion	Criterion
1. Quality of 60 μm flux density	$Q_{60} \geq 2$
2. Range of 60 μm flux density	$f_{60} > 0.6$ Jy
3. Exclusion of stars[a]	$f_{60}^2 > f_{12}f_{25}$
4. Exclusion of hot H II regions and	
cool cirrus[b]	$0.8 < f_{100}/f_{60} < 5.0$

[a] The upper right domain of a dashed line in Fig. 3a.
[b] The zone between two dashed lines in Fig. 3b.

Table 1.   Our infrared flux-density and color criteria

the criterion (iii), $f_{60}{}^2 > f_{12}f_{25}$, following to Strauss et al. (1990). Since most of the galaxies are seen both in $60\mu m$ and $100\mu m$ bands and cooler cirrus sources may be seen only in $100\mu m$, the upper limit of the criteria (iv), $f_{100}/f_{60} < 5.0$, is set to reject the cooler cirrus sources and young stellar objects. The lower limit of (iv), $f_{100}/f_{60} > 0.8$, is set to reject the warmer HII regions as well. The upper cut-off of f_{100}/f_{60} ratio inevitably makes our survey less complete because the at the region where galactic cirrus dominates the upper-limit values refered from IPSC are usually very large. The detailed concepts and procedures of our search is described in Yamada et al. (1993b).

Applying our criteria on *IRAS* point sources in the region of Southern Milky Way between $l = 210°$ and $360°$ at $| b | < 15°$, we selected out 4252 sources (see Table 2).

Category	Numbers[a]
Sources selected by infrared criteria	4252
Identified galaxy candidates	966 (387)
Objects in CGQIRAS[b]	423 (348)
Objects with $f_{60} > 1.936$ Jy	273 (202)

[a] Numbers in parentheses are the numbers of galaxies with measured redshifts.
[b] Numbers of galaxies already cataloged in CGQIRAS (Lonsdale et al. 1989).

Table 2.   Summery of the results of our survey

3.   Large-Scale Structures Crossing the Milky Way

The sky distributions of all the infrared-selected sources, *IRAS* galaxies, and sources not identified as galaxy are shown, respectively, in Figures 1a, 1b, and 1c; It is clear in Figure 1 that we could find only a few galaxies at $l > 260°$ and $| b | < 5°$, where Galactic molecular clouds dominate (Yamada et al. 1993a). In Figures 1b, the sources identified as galaxies are clustered into three high density regions crossing the zone of avoidance nearly vertically around $l \sim 240°$, $280°$, and $315°$. The solid-line histograms in Figure 2 show the projected densities of the *IRAS* galaxies with $f_{60} > 0.6$ Jy detected in our search as functions of the Galactic longitude in three zones of $5° < b < 15°$ (upper panel), $-5° < b < 5°$ (middle panel), and $-15° < b < -5°$ (lower panel). In spite of rather low detectabilities of *IRAS* galaxies in our search, the *IRAS* galaxy densities are higher than the whole sky average in the three regions of longitude both at $5° < b < 15°$ and $-15° < b < -5°$; the peaks are around $l = 245°$, $280°$, and $310°$ at $5° < b < 15°$ and around $l = 240°$, $280°$, and $330°$ at $-15° < b < -5°$. These enhanced density regions have widths of $\sim 10°$ and are located just on the extensions of the three filamentary structures in ESO galaxies toward the Milky Way region (see Figure 1 in Yamada et al. 1993a). In the Puppis filament

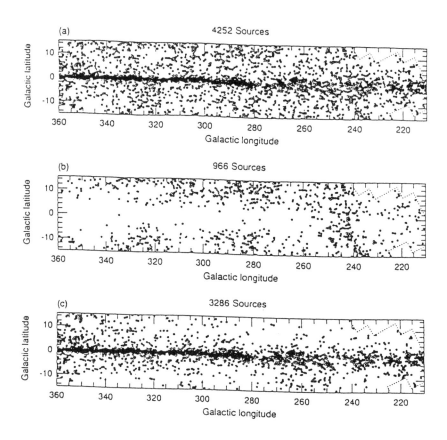

Figure 1.   (a) Sky distributions of all the 4252 *IRAS* sources selected from IPSC with our infrared criteria. (b) The same as (a) but for the 966 sources identified as galaxy. (c) The same as (a) but for the 3256 sources not identified as galaxy.

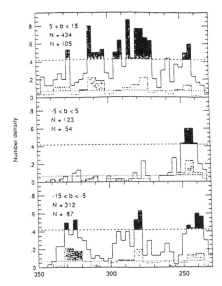

Figure 2.   Histogram of the projected *IRAS* galaxy densities per square degree as functions of Galactic longitude between $b = 5°$ to $15°$ (upper), $-5°$ to $5°$ (middle), and, $-15°$ to $-5°$ (lower). Solid lines are for galaxies with $f_{60} > 0.6$ Jy and dash-dotted lines for galaxies with $f_{60} > 1.936$ Jy.

(crossing at $l \sim 240°$) the galaxy number densities are highest at $\mid b \mid < 5°$ where molecular clouds are little seen (Dame et al. 1987) and the Galactic extinction is known to be the lowest in the whole Milky Way at $\mid b \mid < 5°$. The enhancement is denoted as the Puppis concentration. Since the enhanced density regions in the large-scale distribution of *IRAS* galaxies are also enhanced in optical galaxies (Saunders et al. 1991, Scharf et al. 1992, Strauss et al. 1992a), the three filamentary structures of ESO catalog galaxies are large scale coherent structures running across the Milky Way.

4.   The Puppis Hidden Concentration of Galaxies

4.1.   Motivation

Revealing the galaxy distribution in the zone of avoidance is important in studying the origin of the peculiar motion (i.e., the systematic deviation from the Hubble flow) of the Local Group. It is generally thought that the velocity is induced by the net gravity caused by inhomogeneous distribution of matter around the Local Group. Although this gravitational instability picture seems to be acceptable from the alinement between the observed galaxy dipole and the Local Group velocity vectors, the origin of the Local Group motion is not fully understood yet. It is interesting to divide the Local Group peculiar velocity into three components in the supergalactic cartesian coordinate frame; they are, $(V_{SGX}, V_{SGY}, V_{SGZ})=(-348$ km s^{-1}, 315 km s^{-1}, -373 km s$^{-1})$. The results are not surprising for the velocity components in SGX–SGY plane, since the Virgo cluster, which is the nearest rich cluster from us, exists just along +SGY axis and the Great Attractor region, which is the rechest region on the sky in the distribution of bright galaxies, exists just along the direction of –SGX axis. The two large concentration of galaxies thus might cause significant effects on the Local Group velocity. The clusters themselves and the assembled (infalling) galaxies *around* them must be responsible for the inhomogeniety which induces the Local Group motion.

How about the other component which is *perpendicular* to the supergalactic plane ? It is natural to suppose the existance of other nearby large concentrations of galaxies toward the –SGZ direction. No such concentration of galaxies have previously been known, except the Hydra cluster and the Fornax-Eridanus clusters which are lying below the supergalactic plane and should have *some* contribution to the –SGZ component. However, this could be due to the fact that the directions of ±SGZ axis happens to be inside the zone of avoidance. It is thus interesting to see whether there is any large concentration of galaxies toward the –SGZ direction which corresponds with $(l, b) \sim (227°, -6°)$.

4.2. Puppis Region

Large projected number-density of galaxies in the Puppis region behind the Milky Way ($230° < l < 260°$, $| b | < 15°$) has come to be recognized (Saitō et al. 1990, 1991; Schalf et al. 1992; Yamada et al. 1993a, b). The large extension of the enhanced-density region on the sky and the relatively large diameters of the galaxies concerned indicate the existence of a nearby concentration of galaxies (Kraan-Korteweg and Huchtmeier 1992; Yamada et al. 1993a). Thus the Puppis concentration $[(SGL, SGB) \sim (155°, -75°)]$ is a good candidate for generating such an acceleration towards the –SGZ direction. Lahav et al. (1993) showed rather qualitatively that the Puppis concentration seems to be the most dominant structure on the sky next to the Virgo cluster in the nearby Local Universe within 30 h^{-1}Mpc, using Strauss et al.'s (1990, 1992b) and Yamada et al.'s (1993b) 1.936 Jy samples of *IRAS* galaxies. Kraan-Korteweg & Huchtmeier (1992) measured radial velocities of about 40 galaxies in this region using the HI 21cm line; they found that many of the detected galaxies have radial velocities lower than 3000 km s^{-1} and that some of them are clustered into a group at $cz \sim$ 1500 km s^{-1}. In order to evaluate the significance of the Puppis region, however, a more systematic study of the radial velocity distribution of the galaxies in this zone is needed.

4.3. The Sample of IRAS Galaxies

We construct a far-infrared selected sample of galaxies from the catalog of *IRAS* galaxies given by Yamada et al. (1993b). In the Puppis region defined as $230° < l < 260°$ and $| b | < 15°$, Yamada et al. (1993b) detected 271 *IRAS* galaxies with 60 μm flux-density, $f_{60} > 0.6$ Jy and with optical diameter $\theta \geq 0.11$ arcmin (i.e., 0.1 mm on the UK Schmidt Atlas) in total, 86 of which lie at $-5° < b < 5°$; more than half of the 271 galaxies are newly identified by them. Figure 3 shows the sky distribution of the 271 *IRAS* galaxies; the filled circles are the galaxies whose radial velocities are known so far through this work. Note that many galaxies (75%, i.e. 201 of 271 galaxies) are clustered in the strip with

15° width around $l = 245°$ shown by the two dotted lines, which covers only half of the Puppis region as shown in Figure 3.

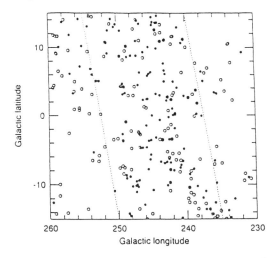

Figure 3.   The sky distribution of the 271 *IRAS* galaxies with $f_{60} >$ 0.6 Jy in the Puppis region. The filled circles are the galaxies whose redshifts are obtained. The open circles with crosses are the galaxies observed but not detected. 75% of the galaxies are distributed in the strip between the two dotted lines and the priority of our redshift observation was given to the galaxies in this zone.

4.4. Radial Velocity Distribution

Figure 4 shows the resultant radial velocity distribution for 97 *IRAS* galaxies with $f_{60} > 0.6$ Jy in the mainly-surveyed strip up to $cz = 12000$ km s^{-1}. A clustering of galaxies around $cz = 2000$ km s^{-1} is evident. The dotted line indicates the smoothed radial velocity distribution of *IRAS* galaxies with $f_{60} >$ 0.6 Jy averaged over the sky at $| b | > 20°$, taken from the QDOT redshift survey by Saunders et al. (1990) and normalized with the area of the strip (450 deg^2), and the dashed line is the same but normalized with the number of galaxies (201) in the surveyed strip.

Even though our radial-velocity information on the flux limited sample is not complete, because both of the incompleteness of the sample and the incompleteness of the redshift survey, we can state with a good degree of confidence that (1) there is a large inhomogeneity in the distribution of *IRAS* galaxies in the direction of the Puppis region and (2) the number density of *IRAS* galaxies at 1500 km s^{-1} $\lesssim cz \lesssim$ 3000 km s^{-1} is at least twice as high as that expected from the whole-sky average. The most prominent peak in Figure 4 between 1600 and 2400 km s^{-1} contains 28 *IRAS* galaxies in the sampled volume of about 420 h^{-3}Mpc^3, while the corresponding whole-sky average value is about ten. Thus the peak density of the Puppis concentration is about 2.8 at this scale $(\sim \sqrt[3]{420} \approx 7.5$ h^{-1}Mpc), although we should note that there are uncertainties in this value because of our binning of 800 km s^{-1}, peculiar velocities of the

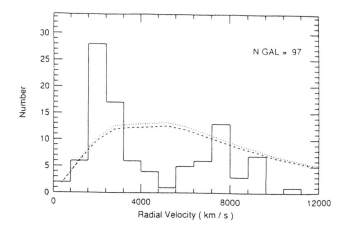

Figure 4.   The radial velocity distribution of the *IRAS* selected galaxies in the strip shown in Figure 3. The width of bins are 800 km s^{-1}.
The dotted line is the average distribution of the QDOT survey at
| b |> 20° by Saunders et al. (1990) which is scaled with the surveyed
area of 450 drg^2 and the dashed line is the same but scaled with the
number of the galaxies (201) in this region.

galaxies, and the shape of our sampling volume; conservatively, we conclude
that the peak density of the Puppis at 7.5 h^{-1}Mpc scale is between twice and
three times of the mean. Saunders et al. (1991) found a new supercluster at
$(l, b)=(220°, -15°)$ and at $cz = 2500$ km s^{-1} in the QDOT redshift survey. The
supercluster (named as S1 by Saunders et al.) is one of the six superclusters
detected within 35 h^{-1}Mpc with a (gaussian) smoothing radius of 5 h^{-1}Mpc in
the QDOT survey. The angular separation between the Puppis concentration
and the S1 supercluster ($\sim 25°$) corresponds to ~ 9 h^{-1}Mpc at the distance of
20 h^{-1}Mpc. Although the relation between the Puppis concentration and the
S1 supercluster in QDOT is not clear so far, it is likely that they form one larger
association. The Puppis concentration may form a dominant part of this association, since the peak of the surface density of the bright *IRAS* galaxies within 30
h^{-1}Mpc lies at $(l, b)\sim(240°, 0°)$ (Lahav et al. 1993). The spacial density peak
of S1 cluster, with 5 h^{-1}Mpc gaussian smoothing radius, was estimated to be
2.63 times the mean (Saunders et al. 1991). Inclusion of the galaxies in the Pup-
pis concentration ,which *were not included* in the QDOT sample at | b |< 10°,
makes this value larger. Note that the peak densities of the Virgo, the Hydra,
the Fornax-Eridanus, and the Centaurus clusters (superclusters) were estimated
to be 4.04, 4.36, 2.48, and 2.98 times the mean density, respectively (Saunders
et al. 1991). Thus the Puppis-S1 association may be one of the largest density
peak in the nearby Local Universe within ~ 50 h^{-1}Mpc.

4.5. Overdensity Profile in 1.936 Jy sample

It is interesting to compare the overdensity of the Puppis concentration more directly with other nearby clusters of galaxies in various scale. However, it is not straightforward to combine our Puppis sample of galaxies with the QDOT sample, because the QDOT sample is constructed by one-in-six sampling whereas the Puppis galaxies are fully sampled but the sample possibly incomplete. For the purpose, therefore, here we use the *IRAS* 1.936 Jy sample by Strauss et al. (1992b) which is a fully sampled catalog at $| \ b \ | > 5°$. The weak-point of using this sample is that the observed number of galaxies in unit volumn is rather small and thus statistical errors are larger than in using a 0.6 Jy sample.

We constructed a list of *IRAS* galaxies with $f_{60} > 1.936$ Jy, following the procedure of Lahav et al. (1993), gathering the data by Strauss et al. (1992b) ($| \ b \ | > 5°$) and by Yamada et al. (1992b) (in the Puppis region; $230° < l < 260°$ and $| \ b \ | < 5°$). The region at $| \ b \ | < 5°$ outside the Puppis region is supplmented by the random distribution of the galaxies assuming the same selection function as Strauss et al.'s survey.

Firstly, we searched fairly large density peaks in the redshift space using the top-hat spheres with 5 h^{-1}Mpc radius centred on the grid points in the region defined by $210° < l < 270°$, $-30° < b < 30°$, and 1000 km s$^{-1} \le cz \le 2500$ km s^{-1}. Points with large overdensity are concentrated into the sky region at $240° < l < 250°$ and $-5° < b < 10°$, and the radial velocity of ~ 2000 km s^{-1}. In Figure 5 we show the overdensity profile centred on $(l, b)=(246°, 2°)$. This

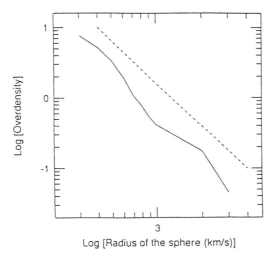

Figure 5.   Density profile around the Puppis nearby concentration in the distribution of the *IRAS* galaxies with $f_{60} > 1.936$ Jy. The quantity plotted is the overdensity in the redshift space, $\delta\rho/\rho$ averaged within concentric spheres centered on the cluster. The diagonal dashed line is a guideline with a slope of -2 arbitrarily normalized, same as in the Figure 9 of Strauss et al. (1992a).

can be directly compared with the overdensities of the Virgo and other clusters studied by Strauss et al. (1992a) using the 1.936 Jy sample (see their Figure

T. Yamada

9). The trend of the fall-off of the overdensity with increasing radius is similar to those of other clusters; we show the arbitrary normalized guideline with the slope of r^{-2} as in Figure 9 of Strauss et al. (1992a). The overdensity of the Puppis concentration within 5 h^{-1}Mpc is as large as those of the Hydra ($\approx$3) and the Ursa Major ($\approx$3) and significantly larger than that of the Fornax ($\approx$1), but less than a half of those of the Virgo ($\approx$7) and the Centaurus ($\approx$8). This is roughly consistent with the result using the 0.6 Jy sample discussed above.

Acknowledgments. I thank the late Satio Hayakawa and his family and Astronomical Society of Japan for allocating me the Hayakawa grants.

References

Dame, T.M., Ungerechts, H., Cohen, R.S., de Geus, E.J., Grenier, I.A., May, J., Murphy, D.C., Nyman, L.-Å., & Thaddeus, P. 1987, ApJ, 322, 706

Kraan-Korteweg, R. C., & Huchtmeier, W.K. 1992, å, 266, 150

Lahav, O., Yamada, T., Schalf, C., & Kraan-Korteweg, R. C. 1993, MNRAS, 262, 711

Saitō, M., Ohtani, H., Asonuma, A., Kashikawa, N., Maki, T., Nishida, S., & Watanabe, T. 1990, PASJ, 42, 603

Saitō, M., Ohtani, H., Baba, A., Hotta, H., Kameno, S., Kurosu, S., Nakada, K., & Takata, T. 1991, PASJ, 43, 449

Saitō, M., Takata, T., & Yamada, T. 1994, PASJ, in press.

Saunders, W., Rowan-Robinson, M., Lawrence, A., Efstathiou, G., Kaiser, N., Ellis, R., Frenk, C.S. 1990, MNRAS, 242, 318

Saunders, W., Frenk, C., Rowan-Robinson, M., Efstathiou, G., Lawrence, A., Kaiser, N., Ellis, R., Crawford, J., Xiao-Yang, X., & Parry, I. 1991, Nature, 349, 32.

Scharf, C., Hoffman, Y., Lahav, O., & Lynden-Bell, D. 1992, MNRAS, 256, 229

Strauss, M. A., Davis, M., Yahil, A., & Huchra, J. P. 1990, ApJ, 361, 49

Strauss, M. A., Davis, M., Yahil, A., & Huchra, J. P. 1992a, ApJ, 385, 421

Strauss, M. A., Huchra, J. P., Davis, M., Yahil, A., Fisher, K. B., & Tonry, J. 1992b, ApJS, 83, 29

Takata, T., Yamada, T., Saitō, M., Kazés, I., & Chamaraux, P. 1994, A&Ap, in press.

Yamada, T., & Saitō, M. 1993, PASJ, 45, 25

Yamada, T., Takata, T., Djamaluddin, T., Tomita, A., Aoki, K., Takeda, A. & Saitō, M. 1993a, MNRAS, 262, 79

Yamada, T., Takata, T., Djamaluddin, T., Tomita, A., Aoki, K., Takeda, A. & Saitō, M. 1993b, ApJS, 84, 59

Yamada, T. Tomita, A., Saitō, M., Chamaraux, P. & Kazés, I. 1994, MNRAS, in press.

Discussion

M. Hauschildt-Purves: Is there a difference in the redshift distribution between the strip defined as the Puppis concentration and the area adjacent to it?

T. Yamada: The number of sources for which we obtained redshifts outside the main strip is rather small. Many of them are found between 1000 and 3000 km/s.

C. Balkowski: Could you give more details about the IR observations you plan to do in the Puppis region?

T. Yamada: There are some regions where IRAS galaxies clump. It is interesting to observe these regions with an IR array to see whether we find concentrations of elliptical galaxies or not.

E. Meurs: One of your future plans is to search for galaxies behind molecular clouds. Does this refer to HI work?

T. Yamada: Yes, we have observed candidate IRAS sources in the Taurus molecular region with the Nançay radiotelescope.

L. Gouguenheim: Because you started from optically identified galaxies, these overdensities are located in transparent regions. It should be interesting to study their extension in surrounding dusty regions. The only possible way is to do blind searches, possibly starting from IRAS sources without optical counterpart. However, from our own experience, the detection rate is rather low.

W. Saunders: The area you have delimited as the core of the Puppis supercluster coincides **exactly** with the region of least 100 μm emission in the entire galactic plane. So the supercluster may well extend further in regions of higher extinction.

T. Yamada: It is possible and interesting.

N. Lu: Comments to the question raised by Will Saunders. There is a possibility that there exists a void next to Puppis (along the plane) at comparable distance because Puppis is seen through a 'Galactic hole'. If this is the case, and since voids play an equal role compared to galaxy concentrations in cosmology, the gravitational effect of Puppis on the LG would be reduced.

T. Yamada: When considering voids, we better consider the overall distribution of galaxies instead of individual voids (on the other hand, you can approximately describe the mass distribution by clusters).

O. Lahav: Given that Puppis explains only ~ 10% of the negative SGZ motion, it may well be that other features hidden behind the ZOA also contribute to this 'local anomaly'. In particular, the 'Local Void' above the Supergalactic Plane may give a 'push' (as hinted e.g. by the analysis of the optical dipole).

T. Yamada: The central region of the Virgo contributes only about 80 km/s (~ 30%) to the SGY component. However, both Virgo and Puppis extend over ~15 Mpc radius. It is therefore possible that the effect of the total Puppis concentration is much larger. However, I agree with you that the Local Void adds a significant component.

Unveiling Large-Scale Structures Behind the Milky Way
ASP Conference Series, Vol. 67, 1994.
C. Balkowski and R. C. Kraan-Korteweg (eds.)

LEDA : The Lyon-Meudon Extragalactic Data Base

N. Durand, L. Bottinelli, L. Gouguenheim
Observatoire de Paris, section de Meudon F-92195 Meudon Cedex

G. Paturel, R. Garnier, M.C. Marthinet, C. Petit,
Observatoire de Lyon, F-69230 Saint Genis Laval

Abstract. The Lyon-Meudon extragalactic database LEDA content is described together with the various outputs and its network distribution.

1. Introduction

LEDA is the oldest Extragalactic Database. It was created in 1983 and has been continuously updated. The main goal is to collect raw measurements coming directly from the observations and to archive them. Mean homogenized data are calculated in the same spirit as those used by de Vaucouleurs and collaborators when publishing the series of Bright Galaxies Catalogues (RC1, RC2, RC3). In fact, RC3 was created using LEDA.

It contains 96,700 galaxies with 325,850 cross-identifications and the measurements of 115,520 morphological types, 32,311 position angles, 177,400 diameters, 215,300 fluxes, 73,400 radial velocities, plus 80,000 additional items.

2. Output and network distribution

LEDA can be reached using the INTERNET network, as follows:
    telnet lmc.univ-lyon1.fr
    login : leda
Two kinds of terminals are supported: text terminal (VT100) or graphic terminal (X11). This last one gives access to the full on-line graphic capabilities. An eight-bit colour screen and a minimum of 4 Mbyte memory are needed.

The result of any request is received via e-mail as ASCII files for text or as postscript files for charts or images. When using an X11 terminal, charts and images can also be displayed directly on the screen.

3. SQL-like language

The LEDA query language is rather similar to the Standard Query Language developed by IBM. Its principle relies on a sentence which describes the request, and whose structure is always the same :
    select (parameters for output) **where** (conditions) **end.**

4. Diagram of menus

The structure of menus in both VT100 and X11 is as follows:

MAIN MENU   Single object   : Explore raw data
                                     : Explore mean data
                Several objects   : Select from a list of names
                                     : SQL-like selection
                2D-informations   : Charts
                                     : Images
                Information   : Instructions for use
                                     : News
                                   : Status
                                   : LEDA's team
                Exit   : Comments

5. Batch mode

It is possible to extract data or plots on a Flamsteed projection, from a list of galaxy names or positions, just by sending this list (one identifier per line) to: LEDAMAIL@LMC.UNIV-LYON1.FR

The"subject" of the mail must be:

LIST : for obtaining main parameters, i.e. coordinates, names, apparent diameter and axis ratio, apparent magnitude and radial velocity.

LISTALL : for obtaining all astrophysical parameters available.

FLAMEQ (or FLAMGA or FLAMSG) for obtaining a postscript file with a Flamsteed equal area projection of your data in respectively equatorial, galactic or supergalactic coordinates.

It is also possible to send a SQL query (one query by line and one query by message) using the "subject": SQL.

6. To know more

Ask for the leaflet "Instruction for use" either to G. Paturel or to N. Durand or pick it up by a direct connection to LEDA.

References

de Vaucouleurs, G., de Vaucouleurs, A., Corwin, H.G. 1976, Second Reference Catalogue of Bright Galaxies, Texas University Press, Austin (RC2)

de Vaucouleurs, G., de Vaucouleurs, A., Corwin, H.G., Buta, R.J., Paturel, P., Fouqué, P. 1991, Third Reference Catalogue of Bright Galaxies, ed. Springer-Verlag New York Inc. (RC3)

Fouqué, P., Durand, N. Bottinelli, L., Gouguenheim, L., Patuel, G. 1992, Monographie de la Base de données extragalactiques, n°3, Lyon

Paturel, G., Fouqué, P., Bottinelli, L., Gouguenheim, L. 1989, Principal Galaxy Catalogue, Monographie de la Base de Données Extragalactiques, n°1, Lyon .

Discussion

G. Mamon: There are a couple of extragalactic databases in the astronomical community, in particular LEDA and NED (the NASA Extragalactic Database). I have used both NED and LEDA and find them to be complementary. To me the main advantage of LEDA is the possibility to obtain a table of the parameters I choose with the selection criteria I choose (using the 'SQL' interface) with one row of data per galaxy.

Unveiling Large-Scale Structures Behind the Milky Way
ASP Conference Series, Vol. 67, 1994.
C. Balkowski and R. C. Kraan-Korteweg (eds.)

Distribution of IRAS Galaxies behind the Milky Way

Tadafumi Takata

Department of Astronomy, Faculty of Science, Kyoto University,
Sakyo-ku, Kyoto 606-01, JAPAN

Abstract. We have performed a flux-limited ($f_{60} \geq 0.6$ Jy) survey of IRAS galaxies behind the whole Milky Way region ($| b | \leq 15°$). For the visual identification, we used Palomar Observatory Sky Survey (POSS) prints for the northern sky, and UK Schmidt I plates and SERC IIIa-J plates for the southern sky. The number of galaxies we have identified on the plates is 2319, of which 1182 have the measured radial velocities. We are now carrying out a redshift survey of these galaxies.

1. Introduction

The Milky Way covers about one-fourth of the whole sky, and because of its severe extinction, the survey of extragalactic objects avoided this region. In order to clarify nearby large scale structures such as filamentary structures crossing the Milky Way and density enhanced regions which will affect the bulk motion of Local group and Local universe hidden behind the Milky Way, it is necessary to search for galaxies behind the Milky Way. Some trials to identify galaxies behind the Milky Way were carried out by Weinberger (1980), Lu et al. (1990), Bottinelli et al. (1992) and Kraan-Korteweg and Huchtmeier (1992), but there is no search covering whole Milky Way region.

For identifying qualitatively high galaxy density region behind the Milky Way, the search for IRAS galaxies is the best way, because they are suggested to trace the distribution of optically selected galaxies and the number is smaller than them (Babul & Postman 1989).

2. Search and Results

We selected IRAS galaxy candidates by applying some infrared color criteria and a flux limit ($f_{60} \geq 0.6$ Jy) on the source in IRAS Point Source Catalog (IRAS PSC) at $| b | \leq 15°$, and inspected their optical images on the plates. Details of the method of our search are described in Yamada et al. (1993). We searched the southern (Yamada et al. 1993) and the northern (Takata et al. 1994a,b) Milky Way using the UK Schmidt Atlas and SERC IIIa-J plates for the southern search and the POSS prints for the northern ones, respectively. A total number of IRAS galaxies identified in our searches is 2319 galaxies, of which 1182 have the measured radial velocities.

The sky distribution of all of the selected galaxies are shown in Figure 1(a) and that of galaxies with cz ≤ 8000 km s^{-1} are shown in 1(b). The six filamentary structures crossing the Milky Way can be seen at $l \sim 90°, 150°, 190°, 240°, 280°$ and $320°$. They are Lyra-Cygnus filament, which may be an extension of Pisces-Perseus supercluster, the edge of Perseus cluster, Gemini-Monoceros filament named by Takata et al. (1994a), Puppis concentration (Saitō et al. 1991; Yamada et al. 1994), Hydra filament, and Supergalactic plane, respectively.

We are now carrying out a redshift survey of these sample galaxies and intend to unveil the large scale structures hidden behind the Milky Way in recent years.

Figure 1(a)

$0 < cz < 8000$ km/s

Figure 1(b)

References

Babul, A. & Postman, M. 1990, ApJ, 359, 280

Bottinelli, L., Durand, N., Fouqué, P. et al. 1992, A&AS, 93, 173

Kraan-Korteweg, R.C. & Huchtmeier, W.K. 1992, A&A, 266, 150

Lu, N.Y., Dow, M.W., Houck, J.R. et al. 1990, ApJ, 357, 388

Saitō, M., Ohtani, H., Baba, A. et al. 1991, PASJ, 43, 449

Takata, T., Yamada, T., Saitō, M. et al. 1994a, A&AS, 104, in press

Takata, T. et al. 1994b, in preperation

Yamada, T., Takata, T., Thomas, D. et al. 1993, ApJS, 89, 57

Yamada, T., Tomita, A., Saitō et al. 1994, MNRAS, in press

Weinberger, R. 1980, A&AS, 40, 20

Part VII

Conclusion

COMMENTS ON THE CONFERENCE

D. LYNDEN-BELL

Institute of Astronomy, University of Cambridge, Cambridge, U.K.

I do not believe in conference summaries but I agreed to comment on contributions to it.

There is enormous fun to be found in cosmography, the charting of chains and aggregates of galaxies both across the sky and in depth. Kraan-Korteweg, Huchra and Saito have all helped to delineate more clearly the crossings of the great chains of galaxies through the murk of the Milky Way. In earlier times, after looking at the distribution of bright galaxies across the Southern Sky (Figure 1) I remarked that its three-toed structure was reminiscent of a dinosaur's foot; however, I have always had some doubt about the way the toes connected to the dinosaurs heel. Is the great Pavo Indus Telescopium band connected to the Centaurus clusters or to the broader band of galaxies more closely associated with the name the Great Attractor? Fairall's early work raised this question. Now these new studies show that there is a big cluster just South of the Galactic Plane and that the primary connection is slantingly across the Milky Way to the Centaurus clusters. They also reveal that both the other toes of the dinosaur indeed continue right through the plane through Puppis and through Antlia respectively.

Qualitative cosmography is full of the fun of finding previously unrecorded galaxies, but any such field must look for its importance in its impact on others. Large-scale structure studies are seriously hampered by uncertainty in the number density of galaxies within 15° of the Galactic Plane. In general terms the studies by Saito et al., who give for galaxies $> 2'$ or 1.9 ESO arcminutes

$$N_{\mathrm{gal}} = 8.5 \exp(-N_{\mathrm{H}}/3.3 \times 10^{21}) \qquad (1)$$

and a variation of diameter with N_{H} of

$$\theta/\theta_0 = \exp - \left(\frac{N_{\mathrm{H}}}{1.37 \times 10^{22}} \right), \qquad (2)$$

are most useful in giving some quantitative data but those of us who wish to know about the change in N_{gal} with galactic longitude, are even more interested in the change in the 8.5 coefficient from place to place.

Those of us interested in explaining the Local Group's motion through the microwave background are naturally very concerned lest there be another Andromeda tugging at the Galaxy, but demurely hidden behind the dust of the Milky Way. Can we really exclude this possibility from current radio, X-ray,

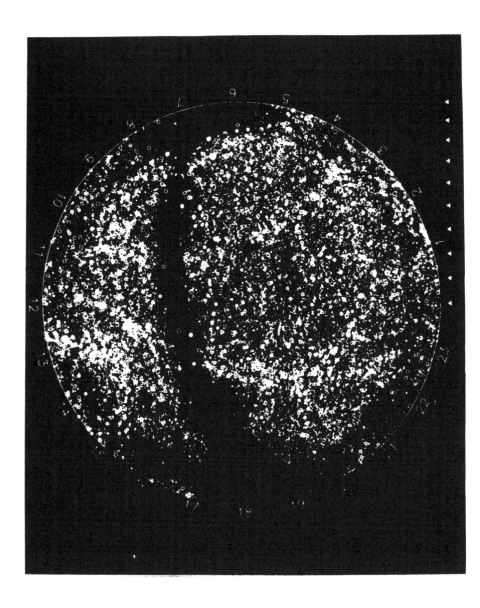

Figure 1.   Equal area projections of southern hemisphere galaxies
(with diameters > 1′) in the ESO/Uppsala Catalogue centred on the
pole.   North of $\delta = -17.5°$ the galaxies come from the MCG except
that north of $\delta = -2.5°$ they come from UGC.

optical or infra-red surveys. While I doubt the existence of any such object, I am less than certain that the possibility can be excluded with certainty. Perhaps we should debate that in the general discussion.

I should emphasize that what we need for Large Scale Structure studies are quantitative studies relating the numbers, magnitudes and sizes of galaxies seen through the obscuration to what would have been seen in its absence. The task is a daunting one and many of the more careful observers have refused to give such uncertain numbers. Unfortunately this makes their results almost unusable for large scale structure purposes. Desperate people have to use even uncertain data and you are the best people to give the estimates, even if you give them large errors. What happens to most scientific papers is something like

Submission – Refereeing – Re-submission – Publication –

10 years on shelves possibly even consulted a few times. After that it is all downhill to a mouldy basement where the volumes gradually decay and form eventually, when damp enough, a good food for those small rapidly moving creatures that the English call silver-fish. If your work is to escape this end it is important that others take it up and build on it. The community that would really like to use your data is the Large Scale Structure community. It is vital for them that the work is accessible and properly calibrated.

From the many papers I heard, those who especially addressed such calibration problems were the studies of Lu, Saunders and Meurs. However, much more calibration of the many surveys is needed. Cameron's paper needs to be followed by using real CCD frames to see how galaxies of each type decrease in diameter under increasing absorption.

John Huchra's fine infrared (2.2μ) CCD frame of Maffei II gave us a foretaste of what the new infrared surveys will achieve by looking through the dust of the Milky Way. Should we all stop and wait because those surveys solve the obscuration problem. For large, bright galaxies I believe this infrared panacea may well be the solution, but at fainter magnitudes there will still be problems. Confusion is already too great at 20° from the galactic plane for large scale star-galaxy separation by the APM. At the lower galactic latitudes the infrared surveys will have yet more sources so the confusion will be far worse than it is now. There are just too many objects. At longitudes near the centre we have had to increase the threshold of the scans, and loose some of the galaxies with $\theta < 1'$ within even 30° of the galactic plane. All this will be worse with the new surveys. To those developing star-galaxy separation, algorithms for the new surveys a good test would be to superpose optical CCD frames at high and low galactic latitudes. By trying to recover the galaxies seen on the high latitude frame one could get a good quantitative measure of the effects of crowding and confusion.

My reason for being somewhat destructive of our work is to get everyone publishing to consider thoroughly the calibration problems and to answer

1) What is the absorption in this field and how great are its fluctuations and our errors in measuring it? How well is it related to N_H?

2) How do the measures of diameter relate to the well known ones of ESO, ESO LV, or UGC diameters?

3) How are the numbers of objects seen as a function of diameter related to the numbers of objects that would have been in UGC as a function of diameter had there been no obscuration?

4) To acknowledge that while redshifts are desirable they are not of much help if the incompleteness due to obscuration is unknown.

One very useful outcome of this conference is the map of the zone of avoidance showing who has been cataloguing faint galaxies in each zone of galactic longitude and latitude.

Finally, if all conferences organised by women scientists have such excellent books describing the programme and such fine sling bags for carrying them, then women scientists will be in grave danger of having to organise all the conferences! You have the heartfelt thanks of all of us.

References outside this volume

Fairall, A.P., 1988, MNRAS, **230**, 69.

Cameron, L.M., 1990, A&A, **233**, 16.

Lynden-Bell, D. & Lahav, O., 1988, *Large Scale Motion in the Universe*, eds. Rubin, V.C. & Coyne, G.V., Princeton UP, p.200.

Discussion

G. Mamon: I would like to iterate the conclusions of my image simulations for the DENIS 2 micron survey. They show that DENIS (and 2 MASS as well) will be able to detect galaxies (with 95% confidence) and separate galaxies from stars (again with 95% confidence) with low false positives **right through the galactic plane** for $l > 45°$ and $l < 315°$ for K<12.5 (a little more than 1 mag brighter than the corresponding limit at the galactic poles). At $l = 0$ we should reach the same completeness for K<12.0 up to $| b | = 3°$. Of course, these are idealized simulations and we will do a little worse with the true images. But on the other hand, we have not yet optimized our algorithms so that we should do a little better. Therefore, I would consider my numbers to be reasonable goals that we should attempt to reach with DENIS.

R. Kraan-Korteweg: I fully agree, that it is very important to improve Cameron's formalism for extinction-corrections of the magnitude and diameters of obscured galaxies. It seems, however, hardly necessary to obtain new CCD-images of a large variety of galaxies to study the obscuration effects. The digitized version of the Surface-Photometry data base of the ESO-Uppsala galaxies provides ample information to investigate these effects in detail.

M. Hauschildt-Purves: There are several extended sources in our Galaxy that are bright at several wavelengths. Can we really exclude that there is a relatively nearby relatively large galaxy hiding that we would not be able to detect?

D. Lynden-Bell: Not at present while there are unidentified X-ray and radio sources, but there is a chance that the Hartmann and Burton survey already has the data for anything within 450 km/s.

H. van Woerden: Could another M31 be hiding in the Zone of Avoidance, waiting to be found through analysis of existing radio survey data? The answer is not easy. M31 has a diameter of several degrees and contains a lot of small continuum sources. These would be very hard to distinguish from the many sources in the Galactic disk. As to 21-cm line radiation, the profile of M31 is 600 km/s wide and the edges are not sharp. The line might be hard to see on a baseline which is not flat on such a wide velocity interval. However, mental reference to our early study of M31 with the Dwingeloo telescope (van de Hulst, Raimond and van Woerden 1957, BAN 14, 1) gives a better answer. The 21-cm line profiles measured locally with a beam of 0°.6 diameter in a galaxy of about 5° diameter often has intensities of several degrees Kelvin and widths of ~ 100 km/s or less. Such lines should be readily detectable in the Hartmann-Burton survey of the northern sky, but a careful search in the databank may be necessary, and, of course, the chance of finding smaller galaxies is much greater. In the southern Milky Way, no similar complete survey is available and the chance for a good find might be greater.

R. Stewart: Given a limited time for HI studies what region of the sky would you give top priority?

D. Lynden-Bell: I would not give time to those who search behind the plane until they have corrected the diameters of the galaxies found for absorption and given us the statistical correction to the numbers found. Once that is done this region of the sky is at least as important as any other. I believe 21-cm would be better than the continuum and no doubt some galaxies in the Zone of Avoidance will be there in the Hartmann-Burton survey.

F. Kerr: We have heard a lot about infrared surveys at this conference, but not so much about HI blind searches, because they are mainly for the future. I want to draw attention to the large amount of valuable information that will come from HI studies, and without the confusion and other problems that the IRAS studies have.

L. Gouguenheim: Donald Lynden-Bell raised the important question of how to discover new galaxies in dusty regions, and in particular how concentrations discovered in rather transparent places do extend in their more dusty surroundings. Blind searches do the job at a price which is: observing time needed for a low detection rate. I am impressed by the results presented by Patricia Henning. Starting from IRAS sources -which have no optical counterparts in dusty places- is another possibility. Our experience shows that a rather large IR luminosity range can be investigated in a distance range up to about 50 Mpc (i.e. about 4000 km/s). However, the detection rate, here again, is low. And the problem is that Program Committees are rather reluctant to allocate observing time for such programmes because they claim that the detection rate is too low ...

N. Lu: Fairly uniform and complete IRAS galaxy samples with redshifts now exist at $|b| > 10°$. A fast way to reach an all-sky uniform galaxy sample is to combine the IRAS samples at $|b| > 5°$ and an HI blind survey at $|b| < 5°$. By studying the correlation between FIR and 21-cm luminosities, one could reach uniformity.

W. Saunders: I wish to draw attention to an interesting coincidence - that the region of sky where constructing IRAS catalogues becomes more difficult (with contamination rising $> 50\%$) coincides with the region where optical extinction rises above 2, making optical work difficult. So I suggest that the highest priority for blind HI work is in this region - about 1.5 steradians total. Obviously the longer the integrations the better but complete area coverage is paramount.

C. Balkowski: Following what Lucienne Gouguenheim said about time allocation committee at Nançay, Renée Kraan-Korteweg and myself have had the same problem at ESO. After a first run we had to wait for two years before the next one. Thanks to MEFOS, which was built by our lab for ESO, we can now observe on guaranteed time.

A. Fairall: One of the main motivations for searches behind the southern Milky Way was the 'Great Attractor'. However, we have heard very little of it during this conference. As one of the 'Seven Samurai' who originated the concept, perhaps you (Donald Lynden-Bell) would comment on your present perspective of the 'Great Attractor'.

D. Lynden-Bell: We originally did our survey without the thought of peculiar velocities. When we found a good enough distance indicator we get Ofer Lahav to plot all galaxies centered on the streaming motion. In this, the remarkable structure of the supergalactic bulge in Centaurus showed up, so we deemed we had discovered the cause of the streaming. However, the latest results show the outward streaming continues beyond the Great Attractor so there is still some mystery to elucidate!

L. Gouguenheim: Not seeing a backside infall into the Great Attractor is not a very strong argument against it, because galaxy distance determinations at these rather large distances are not accurate enough, and suffer from a strong Malmquist bias. Teerikorpi has made some simulations showing that the backside infall can barely be seen with the available data: deeper distance surveys are needed. In other respects, I am rather skeptical about the way the data have been (and can be) corrected for the Malmquist bias.

M. Hendry: Can I add that Rafael Guzman and John Lucey at Durham have recently completed a study using their improved $D_n - \sigma$ relation to probe the back infall, and they find no evidence to confirm its existence. Since their indicator has a smaller intrinsic scatter, they require smaller inhomogeneous Malmquist corrections. Although their result is still preliminary, I expect it will shortly add further fuel to the debate over back infall.

W. Saunders: Would it be useful to have peculiar velocity measurements in the direction of the galactic plane? Because we have heard a lot about HI redshifts and NIR imaging anyway. So we could get the peculiar velocities for free, and make a very attractive observing proposal.

H. van Woerden: Comment to Dr. Lahav. Can we do without detailed observations in the Zone of Avoidance? Maybe we can find the broad characteristics of the sky distribution of galaxies. But watch out! Without direct observations we can never find out what is hidden behind the ZOA. We now know that one of the most luminous radio galaxies and two of the most luminous X-ray clusters lie in the Zone of Avoidance. Who knows what else remains to be discovered? Models never tell us the whole truth.

H. van Woerden: Reports at this conference, and the beautiful summary diagram by Renée Kraan-Korteweg, clearly show that the Zone of Avoidance is rapidly being filled in. I wish to emphasize the great importance of homogeneity of data. In Kraan-Korteweg's ZOA catalogue the surface density of galaxies is higher than that in the ESO survey by Lauberts outside ZOA, because her diameter limit is much smaller. In analyzing the galaxy distributions, the diameters should all be reduced to the same surface brightness level, and these surface brightnesses must all be corrected for extinction. It would also be important to have overlap between the surveys made by different teams in different regions with different techniques.

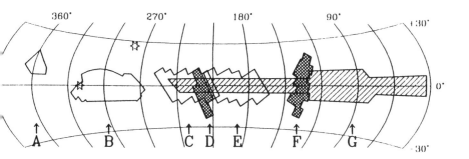

An overview of the optical galaxy searches in the Zone of Avoidance.

Most of these projects are described in these proceedings. Details on the individual searches can be found in the quoted references. The 2 open stars indicate the Microwave Background dipole ($\ell = 280°$, $b = 27°$) and the Great Attractor ($\ell \approx 320°, b \approx 0°$). The individual search areas are:

- A – Ophiuchus SC (Wakamatsu et al., 1994)
    These proceedings

- B – Hydra/Antlia SC and the GA (Kraan-Korteweg and Woudt, 1994)
    These proceedings

- C – Puppis Cluster (Saito et al., 1994)
    These proceedings

- D – Around $\ell \approx 220°$ (Terlevich et al., 1994)
    These proceedings

- E – Neighbourhood of the Perseus-Pisces SC (Pantoja, 1994)
    These proceedings

- F – Crossing of the GP/SGP in the North (Hau, 1994)
    Submitted to A&A.

- G – Whole Northern hemisphere (Seeberger et al., 1994)
    These proceedings

Kraan-Korteweg and Woudt, 1994.

Unveiling Large-Scale Structures Behind the Milky Way
ASP Conference Series, Vol. 67, 1994.
C. Balkowski and R. C. Kraan-Korteweg (eds.)

A Safari through the Zone of Avoidance

At the close of this conference, I would like to express the gratitude of all participants. We all have had a wonderful week in the Zone of Avoidance. We all - a small, privileged band of adventurers. Most colleagues appear to consider this zone at low latitudes, where galaxies are rare, as forbidden territory. 'No Trespassing' - 'Verboden Toegang'. A zone of avoidance for extragalactic astronomers. Few dare to peep through the dusty curtain, like the stumbling explorer on our poster and conference bag. But the sign 'Verboden Toegang' often shields exquisite secrets. It stands at the entrance to the most beautiful nature reserves.

Our experience is that the Zone of Avoidance is a wonderful world, rich in surprises for those who really try to see. There are many galaxies in the ZOA reserve. The density of recorded galaxies is even much higher than outside this zone. The galaxies live in great communities, forming some of the greatest structures ever observed. And within ZOA are unique creatures, like Cygnus A and the X-ray cluster in Ophiuchus. We workshoppers have felt like being on safari. Far from the bustling cities and overcultivated countryside. Not as tourists but as naturalists. Exploring and investigating the population of ZOA. Our safari has been most comfortable. We came upon many excellent waterholes, yielding C_2H_5OH and other aromatic liquors in unlimited quantities. We found great food: tasty plants, birds and game of hitherto unknown varieties. We heard heavenly music at a peaceful site, where big eyes bring the stars close by. And earthly, rhythmic music spurring us to extreme exercise at *'l'Avenir'*. The days in ZOA are long, and the nights are very brief. For after the hardships of exploration and study, there are many relaxing hours of shoptalk and friendship at the campfire near the waterhole ... No safari without guide! Ours proved to be two outstanding experts. In the study of galaxies. In the provision of food and drink. On the steps and scales of music and dance. But above all in the sphere of friendship. With great charm and warmth, they have been good friends to every one of us.

We wish to thank Chantal Balkowski and Renée Kraan-Korteweg for their initiative to hold this workshop, for the excellent arrangements, for opening ZOA to us and showing us its richness and beauty. A bouquet of roses may visualize our gratitude. But I am sure their greater reward will be in our continuing, joint studies in the Zone of Avoidance and in our lasting friendship!

Hugo van Woerden